7.95

Astronomy: An Introduction for the Amateur Astronomer

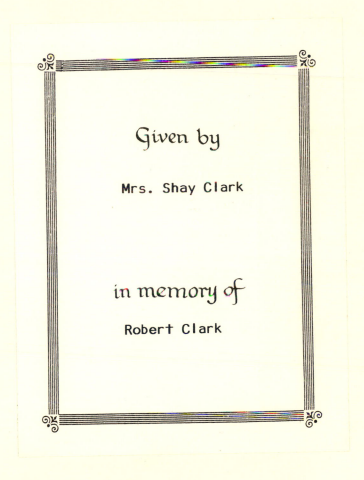

Given by

Mrs. Shay Clark

in memory of

Robert Clark

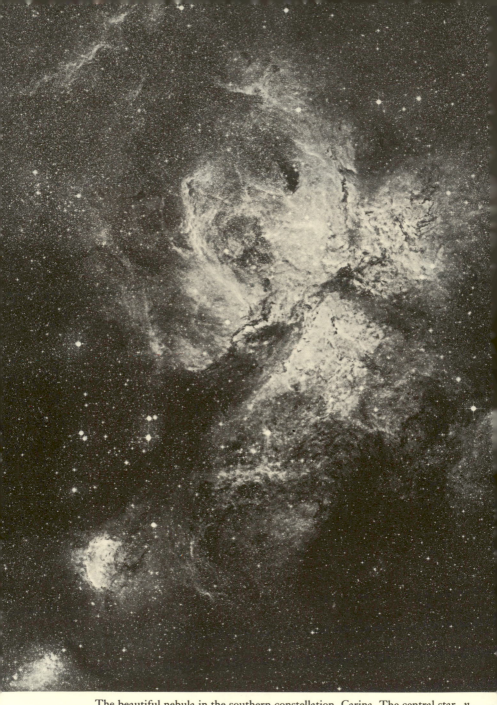

The beautiful nebula in the southern constellation, Carina. The central star, η Carinae, was one of the brightest in the sky in the middle of the nineteenth century, but now it has faded to become invisible to the naked eye

Astronomy: An Introduction for the Amateur Astronomer

JACQUELINE MITTON

CHARLES SCRIBNER'S SONS
New York

Copyright © 1978 Jacqueline Mitton

Library of Congress Cataloging in Publication Data

Mitton, Jacqueline.
  Astronomy.

  Bibliography: p.
    Includes index.
    1. Astronomy. I. Title.
QB45.M693 1979   520   78-13383
ISBN 0-684-16042-0

1 3 5 7 9 11 13 15 17 19   Q/C   20 18 16 14 12 10 8 6 4 2

PRINTED IN THE UNITED STATES OF AMERICA

To Julia and Mary

Contents

		page 11
	Preface	11
1	Introduction	13
2	The Moon	16
3	Eclipses	23
4	Mercury and Venus	27
5	Mars	32
6	Jupiter	35
7	Saturn	39
8	The discovery of the minor planets, Uranus, Neptune and Pluto	41
9	The Earth as a planet	44
10	Comets	48
11	Meteors	53
12	The Earth's motion in space	55
13	The Night Sky	58
14	The paths of the Sun, Moon and planets	64
15	Keeping time	68
16	Kepler and Newton, and the Laws of Gravity	70
17	The Sun	74
18	Matter, light and energy	81
19	Stellar spectra	87
20	The stars—physical properties	90
21	The Hertzsprung-Russell diagram and stellar evolution	94
22	Double stars	101
23	Intrinsic variable stars	105
24	The Milky Way	108
25	The Universe of galaxies	115
26	Telescopes	122
27	Cosmology: the nature of the Universe	129
	Appendices:	
	A Index notation	133
	B Units and abbreviations	134
	C Greek alphabet	135
	D Conversion factors for units of measurement	136
	Suggestions for further reading and reference	137
	Index	139

Illustrations

Photographs

The beautiful nebula in the southern constellation
Carina   *frontispiece*

1. The Moon: a picture constructed from two photographs,
    one taken at first quarter and the other at last
    quarter   *page* 17
2. A close-up of part of the Moon's surface   19
3. The total eclipse of the Sun of 1970 March 7   25
4. A close-up of Mercury photographed by Mariner 10   29
5. A photograph of Venus taken in ultraviolet light   30
6. A mosaic of photographs taken by Viking I showing part
    of the surface of Mars   33
7. Jupiter photographed by the Pioneer 10 spacecraft   36
8. Saturn and ring system   40
9. Comet Ikeya-Seki, 1965, photographed from Mount Wilson
    Observatory, California   49
10. Star trails around the south celestial pole   61
11. A complex group of sunspots   75
12. A solar prominence   77
13. Typical absorption line spectra of stars   88
14. The Crab Nebula   99
15. The Pleiades, showing the reflexion nebulosity   109
16. The Horsehead, a famous dark nebula in the constellation
    of Orion   110
17. The Great Nebula in Orion   111
18. The twin open clusters, h and χ Persei   113
19. The globular cluster ω Centauri   114
20. The spiral galaxy M83   116
21. A distinctively barred spiral galaxy NGC 1365   117
22. The irregular galaxy, M82   118
23. A cluster of galaxies called Abell 1060   120
24. The Anglo-Australian telescope at Siding Spring   125
25. The large radio dish of the Max-Planck Institute for
    radio astronomy, Bonn   127

Text Figures

2.1	Profile of a typical lunar crater	*page* 16
2.2	Prominent features on the Moon	18
2.3	The phases of the Moon	21
2.4	Synodic and sidereal months	21
2.5	The distance of the Moon	22
3.1	Circumstances of solar eclipses	24
3.2	The circumstances of a lunar eclipse	24
4.1	Relative positions of the Earth, Sun, Mercury and Venus	24
4.2	The phases of Mercury or Venus	28
5.1	Oppositions of Mars	32
6.1	Typical motion of Jupiter's moons	37
6.2	The cycle of phenomena of Jupiter's satellites	38
7.1	The orientation of Saturn's rings	39
7.2	Cross-section through Saturn's rings	40
9.1	The structure of the Earth's atmosphere	44
9.2	Interior structure of the Earth	45
9.3	The shape of the Earth's magnetic field	46
9.4	Cross-section through the Van Allen belts	46
10.1	The parts of a comet	48
10.2	Parabolic comet orbit	51
10.3	Elliptical comet orbits	51
11.1	How meteor showers occur	54
12.1	The different star patterns viewed from the Earth's night side during the course of a year	55
12.2	Tilt of the Earth's rotation axis causes the seasons	56
12.3	Dilution of the Sun's rays when the Sun's altitude is low	56
12.4	The Earth's rotation axis and precession	57
13.1	The imaginary celestial sphere	58
13.2	Right ascension and declination on the celestial sphere	59
13.3	The celestial north pole	60
13.4	The observer's hemisphere	62
14.1	The position of the ecliptic on the celestial sphere	64
14.2	The celestial equator and the ecliptic at midday in spring, summer, autumn and winter	65
14.3	Retrograde motion of a superior planet	67
16.1	The geometry of the ellipse	70
16.2	Kepler's second law of planetary motion	70
16.3	Circular orbit	71
17.1	The differential rotation of the Sun	78

17.2	The sunspot cycle	*page* 78
17.3	The solar 'Butterfly Diagram'	78
17.4	General overall structure of the Sun	79
18.1	Definition of wavelength, λ	81
18.2	Dispersion by a glass prism	82
18.3	The electromagnetic spectrum	82
18.4	Planck curves showing the distribution of radiation with wavelength for the continuous spectrum of a black body	85
20.1	Stellar parallax and the definition of the parsec	90
21.1	Schematic version of the Hertzsprung-Russell diagram showing the principal regions in which stars are found	94
21.2	Schematic Hertzsprung-Russell diagram of stars in the solar neighbourhood	95
21.3	Schematic Hertzsprung-Russell diagrams for a young and an old star cluster	97
21.4	Location of the Crab Nebula	98
22.1	The apparent relative motion of the visual binary system α Centauri	101
22.2	Schematic spectrum and radial velocity curves of a double-lined spectroscopic binary	102
22.3	The light curve of the eclipsing binary Algol (β Persei)	103
23.1	Pulsation of a Cepheid variable and corresponding light curve	105
23.2	The light curve of Mira Ceti	106
24.1	The location of Betelgeuse and the Orion Nebula (M42)	112
24.2	The distribution of globular clusters in our Galaxy	114
25.1	Hubble's scheme for the classification of galaxies	115
25.2	Hubble's Law	119
26.1	Beams of parallel light brought to foci by a convex lens and a concave mirror	122
26.2	The various arrangements for reflecting telescopes	123
26.3	The angular magnification of a telescope	124
26.4	The meaning of resolution	124
26.5	The construction of an achromatic doublet	125
26.6	Spherical aberration	126

Preface

This book grew out of notes prepared for a one-year school course and evening classes in elementary astronomy given on behalf of the University of Cambridge Board of Extra-Mural Studies. It is intended not only for students, however, but for anyone who is interested in finding out what astronomy today is about. I have deliberately been concise in the text to leave room for the illustrations that often say more than words can. The emphasis is on the science behind astronomy, and current astronomical knowledge, rather than the practical aspects of amateur astronomy.

Throughout, the metric system of units is used where numerical measurements are mentioned. These are the units actually used by scientists in their work, but some readers may not find them as familiar as the feet and pounds used for everyday measurement. Appendix D gives the main conversion factors to help you.

My grateful thanks go to Derek McNally for many helpful comments on the original manuscript, and to my husband, Simon, for comments and contributions, especially Section 27. Special thanks are due to my mother for her patient and accurate typing of the manuscript.

Jacqueline Mitton
Institute of Astronomy
University of Cambridge

1 *Introduction*

One of the most awe-inspiring sights in nature is the myriads of stars burning in a clear, dark sky. As far back in time as historical and archaeological records can take us, men have been gripped by the fascination of astronomy. By ingenuity and perseverance through the centuries, from mere specks of light in the heavens, the true scale and nature of the Universe has gradually been unfolded. The story of astronomy's history is in itself of great interest, but in this book I shall concentrate on presenting an up-to-date view of the Universe and some of the basic laws of nature which make the Universe how it is. Of course, nobody can give a complete account because there are still plenty of unanswered questions and, all the time, improved technology is extending the limits of the observable Universe. Neither can astronomy hope to answer the question 'Why does the Universe exist?', even if it strives to understand the natural laws which govern it. However, if you have been intrigued by the beauty of the heavens, or the nature of your environment—in the broadest sense—you will be curious to know more about those specks of light you can see in the sky, and the many more mysteries which the telescope reveals. So, read on.

The scale of the Universe and the Earth's place in it

The Earth and all life on it depend on the Sun for their existence. The Earth is one of a family of nine planets which are in orbit around the Sun and are the chief members of the solar System. The Sun is a typical star. Planets and stars differ fundamentally in that stars are power-houses sending out vast amounts of heat and light energy, whereas planets are intrinsically cold by comparison. Planets have no ongoing source of energy, primarily because they have too little matter in them to become like stars. The planets shine only by the sunlight which they reflect.

Our nearest neighbour in space is the Moon, which orbits the Earth at a distance of $3 \cdot 8 \times 10^7$ m*. The mean distance between the Earth and the Sun, $1 \cdot 5 \times 10^{11}$ m, is called the **Astronomical Unit** (A.U.); it is often used as a unit of measurement for distances in the Solar System.

Compared with the other members of the Solar System, even the

---

*See appendix for explanation of index notation

nearest stars are vastly distant. Our nearest star, called Proxima Centauri, is at a distance of $4 \cdot 1 \times 10^{16}$m. For such large distances, the unit of distance called the **light year** is quite often used. This is simply the distance that light travels (at 3×10^8m/s) in a year, and it amounts to $9 \cdot 5 \times 10^{15}$m. Proxima Centauri is $4 \cdot 3$ light years away. It is impossible to tell whether any other stars have planetary systems, except by very indirect ways which so far have not been conclusive. However, considering the vast numbers of stars there are, it is quite likely that there are planets round some other stars.

All the stars in the night sky which can be seen with the naked eye are comparatively near to us. Stars are not distributed through the whole of space. Rather, all the stars belong to a great congregation called a galaxy. Our Galaxy is shaped somewhat like a flat disc with a large lump in the middle, and from inside we see the disc of the Galaxy as the band of light across the sky that we call the Milky Way. Our Galaxy contains 100,000,000,000 stars and has a diameter of at least 100,000 light years.

However, the Universe does not end at the edge of the Milky Way. As well as stars and planets, faint, misty patches, or nebulae can also be detected in the sky. About a hundred of the brightest of these were catalogued in 1781 by the Frenchman Messier, and are designated M1, M2, etc. Many more have been discovered since. Some of these turned out to be clouds of gas or groups of stars in our own Galaxy, but others are more distant galaxies. The nearest large galaxy, M31 in Andromeda, is at a distance of 2,300,000 light years. The further into space that telescopes probe, the more galaxies are discovered. So we see that the Earth is very insignificant in the Universe as a whole.

TABLE 1.1 The Solar System

planet	diameter (m)	mean distance from Sun (A.U.)	orbital period (yr)
Mercury	$4 \cdot 9 \times 10^6$	0·39	0·24
Venus	$1 \cdot 2 \times 10^7$	0·72	0·62
Earth	$1 \cdot 3 \times 10^7$	1·00	1·00
Mars	$6 \cdot 8 \times 10^6$	1·5	1·9
Jupiter	$1 \cdot 4 \times 10^8$	5·2	11·9
Saturn	$1 \cdot 2 \times 10^8$	9·5	29·5
Uranus	$4 \cdot 7 \times 10^7$	19·2	84
Neptune	$4 \cdot 8 \times 10^7$	30·1	165
Pluto	$5 \cdot 9 \times 10^6$	39·4	248

TABLE 1.2 The distances and apparent magnitudes
of the nearest stars

As some of these stars are very faint they have no proper names, and are referred to only by catalogue numbers.

star	distance (light years)	apparent magnitude
Proxima Centauri	4·3	10·7
α Centauri	4·3	0·0
Barnard's star	6·0	9·5
Wolf 359	7·8	13·7
Lal 21185	8·2	7·5
Sirius	8·7	−1·5
		12·4
UV Ceti	8·7	12·9*
Ross 154	9·4	10·6
Ross 248	10·3	12·3
ε Eridani	10·8	3·7

*binary

2 *The Moon*

The Moon is our nearest neighbour in space, and a companion to our planet. It is often said that the Earth and Moon form a twin-planet; none of the other small planets, apart from Mars, have any natural satellites, and the moons of Mars are extremely tiny—probably captured minor planets. By contrast, the Moon has a diameter of about one-quarter the Earth's.

Of course, a great deal of money and effort has been directed into the exploration of the Moon by both unmanned and manned spacecraft, so more is known about the Moon than any other astronomical body. In fact, it will probably be many years before all the data that has been collected will have been properly examined.

The Moon's surface

The Moon, like all the planets, shines only by the sunlight which it reflects. Observation of the Moon with binoculars or even with the naked eye, shows that some areas are lighter than average, whereas other areas are darker. The light areas are mainly mountainous or upland zones. The large dark areas were termed **maria** (the plural of the Latin word, mare, meaning 'sea') before anything about their true nature was understood. They do not contain any water, for there does not seem to be a drop of water anywhere on the Moon; they are in fact solidified seas of lava left over from a time when the Moon was volcanically active.

The entire greyish surface of the Moon is peppered by countless **craters**, although they are not uniformly distributed, there being fewer in the mare areas. Craters range in size from a maximum of about 200 kilometres diameter down to the smallest distinguishable—only centimetres across. In general, the outer walls of the larger craters rise gently above the surrounding area, but fall steeply on the interior to a level below that of the surrounding plain. The inner walls often show evidence of landslips or terracing. The walls contain enough material to account for

2.1 Profile of a typical lunar crater

1. The Moon. This picture has been constructed from two photographs, one taken at first quarter and the other at last quarter. Careful inspection reveals that the shadows in the two halves point in opposite directions. The photograph is done in this way so that the features are shown up well by their shadows. A photograph of the real full Moon shows comparatively little contrast as the shadows are short then. (Lick Observatory)

that missing from the interior. Frequently the craters have small peaks at their centres. Not all the craters were formed at the same time. There are those that appear to be old and broken down, others that are flooded with lava, overlapping craters, crater-chains and fresh, young-looking craters. Some craters, notably Tycho and Copernicus, have extensive, bright, ray systems spreading out from them. These may be ejected material, or material below the lunar surface layer which was revealed when ejecta disturbed it.

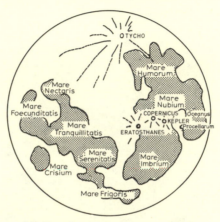

2.2 Prominent features on the Moon

The origin of the lunar craters has been the subject of considerable controversy in the past, but it is known now that the impacts of meteors and not volcanic action are responsible for most craters. There is considerable independent evidence for volcanic activity. Quite a number of lunar specimens similar to terrestrial volcanic lava have been recovered. There are occasional reports of temporary red patches, or obscurations which could be gas clouds, particularly along the borders of the maria, which could be weak spots in the lunar crust. These events are called 'Transient Lunar Phenomena' (or TLP for short). It seems probable that the Moon had many active volcanoes until 3,500 million years ago.

On the flat maria especially there are numerous **rilles** or deep winding valleys, such as the Hadley Rille which was explored by the Apollo 15 astronauts. The origin of these rilles is not certain but they may be the remains of ancient channels along which lava flowed. The Moon also has some impressive mountain ranges, such as the lunar Alps and Apennines, which are comparable in height with their counterparts on Earth. The mountain ranges tend to occur near the borders of the maria.

2. A close-up of part of the Moon's surface. Notice the contrast between the bright rayed crater and the dark interior of the crater just below it in the photograph. (NASA Apollo)

The best time to view lunar features is when they lie close to the **terminator**, the join between the illuminated and dark portions of the Moon. Along the terminator, craters and mountains cast strong shadows and can be seen clearly in relief even with binoculars or a small telescope. It is often difficult to pick out features on a full Moon, because of the lack of contrast, although this is the best time to see the rayed craters.

The Moon's structure and origin

The Moon's mass is only one-eightieth the mass of the Earth and the gravitational pull at the lunar surface is only one-sixth of that at the surface of the Earth. The Moon's gravity is not strong enough to retain any atmosphere at all. This is because gas molecules are in constant motion, and without a strong gravitational attraction, any atmosphere will soon disperse into space. With no atmosphere, the lunar surface suffers no erosion by weather as the Earth does. However, lunar rocks show signs of erosion. This is due to the impacts of microscopic meteorites, and the effect of the sharp changes in temperature between lunar day and night, which cause considerable expansion and contraction in the surface.

Lunar rocks are similar to, but nevertheless significantly different from terrestrial rocks. If the Moon was ever part of the Earth it must have broken away very early in its history to have different rocks. It is too large to be a captured minor planet. It may well be that the Earth and Moon were formed together as a double planet about 4,650 million years ago.

The study of natural and artificially created 'moonquakes' and the lack of any magnetic field show that the Moon does not have a metallic core like the Earth, but distinct discontinuities in the Moon's interior structure have been identified. Instruments left by Apollo astronauts signal records of moonquakes to Earth.

The motion of the Moon

As the Moon orbits around the Earth, the change in the relative positions of the Moon, Earth and Sun cause the Moon to show its phases. The time between consecutive new Moons is 29d 12h and is called the **synodic month**. During this time, however, the Earth, and consequently the Moon's orbit, have travelled some way around the Sun, so the position of the Moon against the background of stars is different. The time for the Moon to return to the same position in the sky as viewed from Earth is called the **sidereal month** and is about two days shorter than the synodic month.

The Moon rotates on its own axis in exactly the time it takes to travel round the Earth so that it always keeps the same face to the Earth. This state of affairs arises from tidal interaction between the Earth and Moon

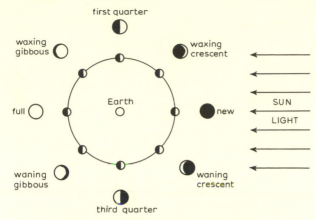

2.3 The phases of the Moon, showing the appearance of the Moon from Earth at different points in its orbit

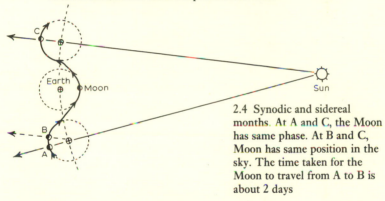

2.4 Synodic and sidereal months. At A and C, the Moon has same phase. At B and C, Moon has same position in the sky. The time taken for the Moon to travel from A to B is about 2 days

that acts like friction to slow down the Moon's rotation over a long period of time. The Moon's orbit is almost circular, and is inclined at 5° to the plane in which the Earth orbits the Sun (the ecliptic). Due to the deviation of the Moon's orbit from an exact circle, and small perturbations on its motion, in all, about 59 per cent of the lunar surface becomes visible from the Earth at some time. This phenomenon is called **libration**. Orbiting spacecraft have shown that the back of the Moon is broadly similar to the Earth-facing side, except that it has no major maria.

One way in which the Moon's motion can be checked accurately is through the timing of **lunar occultations** of stars. The Moon quite often passes over stars whose positions are well known. The times at which these stars disappear and reappear can be observed with great accuracy. Lunar occultations have also been extremely important in accurate measurements of the positions of radio sources in the sky. Unfortunately, the limitations of radio telescopes do not permit the location of radio sources with the same accuracy that optical telescopes can achieve. (See Section 26.)

The measurement of the Moon's distance

Very accurate distance measurements can be made by radar techniques. The time taken for a radar pulse to travel to the Moon and to be reflected back is recorded. As the pulse travels with the speed of light, whose value is known with great precision, the distance can be deduced.

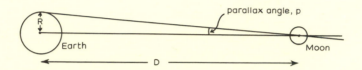

2.5 The distance of the Moon by the parallax method

$$D = \frac{R}{P}$$

Originally, the distance of the Moon was found by the parallax method. The apparent position of the Moon against the background of stars is different from different places on the Earth. The difference in position as viewed from the pole and the equator is usually taken as the standard parallax measurement.

During the Apollo missions small reflectors, rather like those on a car or bicycle were deliberately left on the Moon. By aiming extremely powerful lasers at these reflectors and timing the passage of pulses of laser light from Earth to Moon and back again, scientists can find the Moon's position to within a centimetre.

3 Eclipses

The Moon's orbit around the Earth lies close to the plane of the Earth's orbit around the Sun. The two orbits are inclined to each other at only 5°. As a consequence, it quite often happens that all three bodies are aligned. Under these circumstances an **eclipse** occurs.

Solar Eclipses

The Sun is, of course, much further away and much larger than the Moon, so it is entirely accidental that both appear about the same size in our sky (about $\frac{1}{2}$° across). Only this coincidence makes the spectacle of a total eclipse of the Sun possible. As the Moon's shadow is barely a few miles across, even under the most favourable conditions, totality can only be seen in the narrow band which the shadow sweeps out on the Earth's surface. However, a partial eclipse can be seen over a much wider area. The total duration of the eclipse from start to finish may be several hours, but totality lasts a maximum of 7 minutes and is often shorter. As totality approaches, the last glimpse of the Sun's disc is through the valleys on the Moon's limb, giving the impression of a diamond ring, or a string of pearls. This phenomenon is sometimes called 'Baily's Beads'. The few minutes of totality give astronomers a unique opportunity to study the faint solar corona, the tenuous outer regions of the Sun, which normally are not visible against the brilliance of the photosphere. This is also a time when faint objects close to the Sun might be detected. More than once a new comet has been found near to the Sun during an eclipse.

On average, a total eclipse of the Sun is visible somewhere on Earth every one or two years. It can only happen at new Moon, when the Moon is between the Earth and the Sun, but it does not occur every month because of the inclination of the Moon's orbit. Most months, the Moon's shadow misses the Earth completely. Occasionally, the circumstance for an eclipse may be right, except that the Moon is further away than average and/or the Earth is closer to the Sun than average, so that the apparent diameter of the Moon is less than the Sun's. Then, an **annular eclipse** is seen, so-called because a ring or annulus of the Sun remains visible. Unfortunately, such eclipses are of very little scientific use, as the corona does not become visible, and too much light remains.

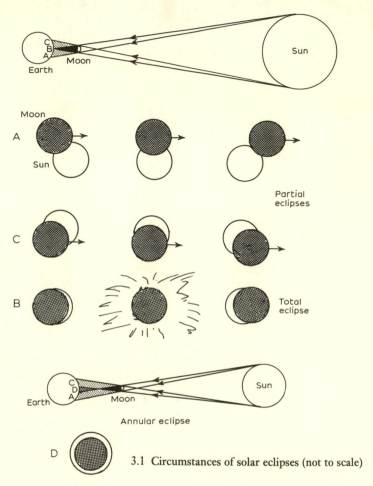

3.1 Circumstances of solar eclipses (not to scale)

Lunar Eclipses

Eclipses of the Moon are a much more frequent occurrence than solar eclipses at any given place. The Earth's shadow is so much larger than the Moon's, that the lunar eclipse can be seen from as much as half the Earth's surface, and totality can last up to about two hours. During totality, the Moon does not vanish from sight completely, but usually looks coppery-red in colour. This happens because the Moon's surface is faintly illuminated by sunlight scattered from the Earth's atmosphere, and red light in preference to blue is scattered towards the Moon.

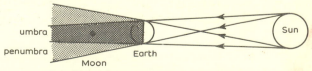

3.2 The circumstances of a lunar eclipse (not to scale)

3. The total eclipse of the Sun of 1970 March 7. A special filter was used to dim the bright inner part of the corona and show up the faint outer parts. (High Altitude Observatory, Colorado)

TABLE 3.1 Forthcoming total solar eclipses

date	area of visibility of total eclipse
1979 February 26	W. coast of N. America
1980 February 16	Central Atlantic
1981 July 31	Black Sea
1983 June 11	Indian Ocean
1984 November 22	New Guinea
1985 November 12	S. Pacific
1986 October 3	W. of Iceland
1988 March 18	Indian Ocean
1990 July 22	Baltic

TABLE 3.2 Forthcoming lunar eclipses

date	extent
1978 September 16	total
1979 March 13	partial
1979 September 6	total
1981 July 17	partial
1982 January 9	total
1982 July 6	total
1982 December 30	total
1983 June 25	partial
1985 May 4	total
1985 October 28	total
1986 April 24	total
1986 October 17	total

4 *Mercury and Venus*

Mercury and Venus are described as the **Inferior Planets** because their orbits lie inside the Earth's orbit. This fact has two important consequences on the appearance of these planets from Earth:

they show **phases** (like the Moon's)

they can be seen only in a limited region of the sky near to the Sun, and so can be observed by the naked eye or small telescopes only just after sunset or just before sunrise.

At **greatest elongation** the planet is at its maximum angular separation from the Sun in the sky. For Mercury this is only 28° and is 47° for Venus. At this point the planet shows half phase or **dichotomy**. At eastern elongation the planet is following the Sun and so is visible after sunset as an evening star. At western elongation the planet is visible before

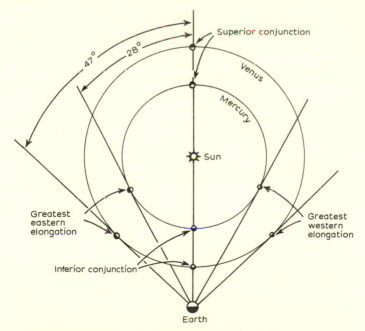

4.1 The chief relative positions of the Earth, the Sun and the inferior planets, Mercury and Venus

sunrise as a morning star. Mercury orbits the Sun so rapidly that it is visible only for a few days at a time, on four or five occasions each year. Even then it is quite difficult to spot.

4.2 The phases of Mercury or Venus as seen from the Earth

Greatest brilliancy occurs at crescent phase. This may seem strange since only a small proportion of the illuminated surface is visible. However, the planet is almost at closest approach to the Earth and therefore appears larger. The increase in angular size has a greater effect on the brightness than the reduction in phase. Venus becomes the brightest object in the sky after the Sun and Moon, reaching a magnitude of −4·4. Mercury's maximum magnitude is −2·3.

Transits

A **transit** is the passage of Venus or Mercury directly between the Earth and the Sun, such that the disc of the planet can be seen as a black spot against the Sun. The timing of transits was of immense historical importance for determining distances in the Solar System before the invention of radar. For example, the distance of Venus can be determined by timing the start and finish of a transit as observed at several different locations on the Earth. Transits of Venus occur in pairs, separated by 8 years, at alternate intervals of 113 and 130 years. The next transit of Venus is in 2004. Transits of Mercury occur on average about every 10 years. The next is in 1986.

Surface conditions on Mercury

Mercury is a small planet, only slightly larger than the Moon. Like the Moon, it reflects only about 10 per cent of the sunlight falling on it. This suggests that Mercury's surface is made of dark volcanic rocks and lava. In 1974–5, the space probe Mariner 10 sent back over ten thousand photographs showing that Mercury's surface is heavily cratered like the Moon's. It used to be thought that Mercury always kept the same face to the Sun, but this was an unfortunate result of the difficulties of observing this tiny planet. Radar measurements have now confirmed that Mercury rotates on its own axis in 58·6 days, while its orbital period is 88 days. The temperature ranges from 350°C at midday to −250°C at night.

4. A close-up of Mercury. The cratered surface bears a striking resemblance to the Moon. (NASA Mariner 10)

Prior to the fly-by of Mariner 10, the existence of an atmosphere around Mercury was in doubt, although observers using large telescopes had reported seeing whitish clouds from time to time. It was difficult to understand how a small, hot planet, with such a low gravitational pull, could retain any atmosphere. However, Mariner 10 confirmed that there is a very tenuous atmosphere, with a most unusual composition. Helium, argon and neon have been detected. The only possible explanations are that these gases are the products of the radioactive decay of rocks on Mercury, or that they have been captured from the streams of particles leaving the Sun. This would explain the presence of an atmosphere on such a small planet, for the gases can be continually replaced as they gradually escape into space. Mercury also has a weak magnetic field.

5. A photograph of Venus taken in ultraviolet light reveals structure in the dense cloud layers. (NASA Mariner 10)

Surface conditions on Venus

Venus is similar in size to the Earth. Telescopic observations reveal no permanent features because the surface is covered by a uniformly thick layer of opaque cloud. Various shadings, and effects due to the scattering of light by the clouds have been reported. Although the planet is veiled by dense clouds, information on the atmospheric composition and surface temperature has been sent back by various space probes. The atmosphere is very dense with a pressure at the surface approaching a hundred times that of the Earth's atmosphere. It is composed chiefly of carbon dioxide with a small amount of nitrogen and traces of inert gases. There seems to be little oxygen or water vapour. The surface temperature is around 400°C. This is much higher than might be expected considering Venus's distance from the Sun. A possible explanation is that the clouds produce what is called a 'greenhouse effect': they trap heat because the carbon dioxide gas is transparent to short-wave infrared waves from the Sun, but opaque to the longer waves produced at the warm surface of the planet.

No magnetic field has been detected around Venus. The period of axial rotation has been found by radar techniques and is about 250 days— slightly longer than Venus's orbital period of 225 days. Furthermore, the rotation is in the opposite direction to the Earth's. If the Sun could be seen through the clouds on Venus, it would move very slowly from west to east in the sky. Radar has also been applied in attempts to make maps of the relief on the surface. This has led to the discovery of several large craters, mountain ranges and plains. In 1975 the Russian spacecraft Venera 9 and Venera 10 succeeded in landing on Venus and sent back pictures of a stark, rocky landscape.

5 *Mars*

In a small telescope, Mars appears as a rust-red disc, with permanent darker markings. There are conspicuous white polar caps which shrink and grow with the seasons. Near the poles especially, there are seasonal variations in the appearance of the surface. Mars' rotation period and the inclination of its axis are similar to the Earth's, so that the martian days and seasons are similar to our own. Mars, however, is only just over half the diameter of the Earth.

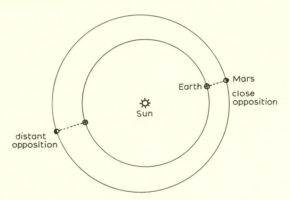

5.1 Oppositions of Mars

The surface conditions

In 1976, United States scientists succeeded in landing their two Viking probes on Mars. These spacecraft were able to do remotely controlled experiments on the physical state and composition of Mars' soil and atmosphere, as well as detailed photography. In addition, the Viking Orbiters returned many photographs of Mars' surface, improving on the results from the earlier Mariner missions.

Like the Moon and Mercury, Mars is speckled with craters, but many of them show evidence of erosion. This is probably the result of the severe dust storms which are raised by the high winds on Mars. These occur particularly at Mars' closest approach to the Sun, when solar heating stirs up tremendous gales on a global scale. When Mariner 9 first arrived in

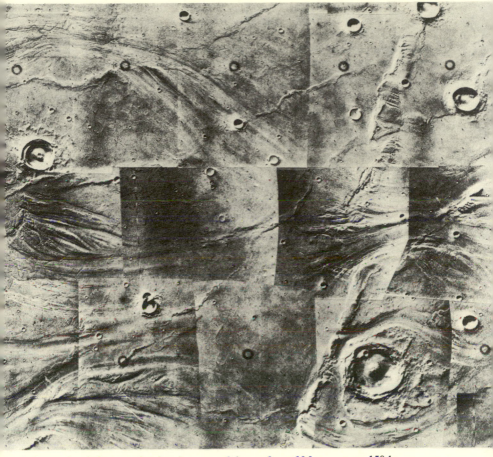

6. A mosaic of photographs showing part of the surface of Mars, an area 150 by 200 km. Lava flows and sinuous, dried-up river channels can be seen as well as many craters. (NASA Viking 1)

1971 the surface was obscured by a mighty dust storm which gradually settled. The martian air looks red on Viking photographs because of all the red dust it holds. In some places, the dusty soil is piled into great dunes.

Mars was once volcanically active. There are several gigantic extinct volcanoes. The largest is known as Olympus Mons. The analysis of martian soil also showed the presence of volcanic lava.

One thing is quite clear: there is no evidence for the long, straight canals which some early observers thought they saw. There are many sinuous valleys that leave little doubt that water once flowed on Mars, though there is no free water now. What water there is lies trapped in Mars' polar ice caps. The permanent ice caps are too warm to be anything but frozen water, but the frost that forms in winter and disappears in summer is frozen carbon dioxide (dry ice).

An outstanding feature on Mars is the great canyon called Valles Marineris. It runs east-west close to the planet's equator, extending for 5,000 km. It is thought to be a huge rift valley roughly 150 km wide and 2 or 3 km deep.

The Viking landers carried out several experiments designed to detect any living organisms that might be present in the soil. Unfortunately, the results were inconclusive. However, it is obvious that very primitive, microscopic organisms are the only possible form of life on Mars. The conditions are not particularly hospitable to life. The atmosphere is 95% carbon dioxide with only traces of oxygen and water. Its pressure is only about 7 millibars compared with an average 1000 millibars on Earth. The temperature rarely rises above the freezing point of water and frequently falls to $-100°C$. Mars is a cold, dry world.

The satellites of Mars

Mars has two natural satellites. Phobos, the inner one, travels around Mars more than three times in each martian day. An observer on Mars would see Phobos rise in the west, and stay risen for about three hours, during which time it would undergo most of its cycle of phases! Mariner 9 photographs of Phobos show that it is an irregularly-shaped lump of rock about 20 km across and pitted with about 100 meteor craters. This suggests that at least some of the craters on Mars are of meteoritic origin.

Deimos, the other satellite of Mars, is smaller than Phobos. Its disc would not be discernible to an observer on Mars.

6 Jupiter

Jupiter is the largest planet in the Solar System. Its diameter is ten times that of the Earth. Its mean density is only 1,300 kg/m³, just greater than the density of water. Although it is a giant planet, Jupiter rotates on its axis once in just under 10 hours, making it the planet with the fastest rotation. Its orbital period is about 12 years.

Telescopic appearance

When viewed through a telescope, Jupiter appears as a yellowish disc. A series of darker bands cross the surface, parallel to the equator. These bands vary in their dimensions and positions from time to time, and also contain irregularities which change fairly quickly. The most intriguing and conspicuous feature is the Great Red Spot. This atmospheric feature is 48,000 km long and 11,000 km wide with a roughly oval shape. Since its discovery in the seventeenth century it has changed colour through various shades of red and pink, and has shifted its positon relative to other features. In fact, the period of rotation of the Great Red Spot has diminished by about 5 seconds since its discovery. Its origin remains mysterious, although observations made by the spacecraft Pioneer 10 in 1974 suggest that it must be some kind of atmospheric phenomenon, probably a gigantic whirlwind. The various bands do not rotate as a solid body. The rotation speed is 6 minutes longer at the poles than it is at the equator. Furthermore, the rotation speed falls off in an irregular way between the equator and poles. The rapid rotation makes Jupiter slightly flattened at its poles.

Physical conditions

The solid surface is covered with frozen gases, so there is no distinct boundary between a solid surface and an atmosphere. The pull of gravity is so strong that even the lightest gases are still held in Jupiter's atmosphere which contains hydrogen, helium, water, methane and ammonia. The temperature is around −146°C.

The light bands on Jupiter are wind circulation belts in which clouds are rising up from the interior of the atmosphere. In contrast, the conspicuous dark belts are zones of descending gas. According to data

7. Jupiter photographed by the Pioneer 10 spacecraft from a distance of 2,500,000 km. The great red spot and the cloud belts show up clearly. The dark spot is the shadow of one of Jupiter's moons, Io. (NASA Pioneer 10)

obtained by the Pioneer craft in 1974, the global circulation pattern is a direct consequence of Jupiter's rapid rotation and the fact that the planet gives out more heat than it receives from the Sun. This extra heat is probably derived by continuous but very slight contraction of Jupiter's dense interior.

The interior of Jupiter must be under exceedingly high pressure due to the great mass of the planet. As the average density is so low, it is thought possible that the interior is composed of hydrogen that has been compressed so much that it is solid. Solid hydrogen would have the properties of a metal, and a metal core inside would explain why there is around Jupiter a gigantic magnetic field stretching a million kilometres into interplanetary space. Electrons trapped by this giant magnetic cage beam out strong radio signals which can be picked up on Earth, even by small radio telescopes.

The satellites of Jupiter

Jupiter has at least 13 known natural satellites, each designated by a roman numeral. The four largest, which are all about the size of Mercury, were discovered by Galileo with one of the first telescopes ever turned on the sky, so they are known collectively as the Galilean Moons. Their individual names are Io (I), Europa (II), Ganymede (III), and Callisto (IV). The other nine satellites are much fainter and can only be seen with large telescopes. Some may be minor planets captured by Jupiter's massive gravitational pull.

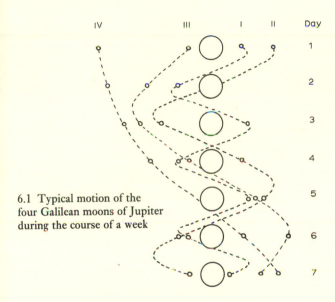

6.1 Typical motion of the four Galilean moons of Jupiter during the course of a week

The motion of the Galilean moons of Jupiter can easily be observed from day to day with field glasses or the smallest of telescopes. As they orbit about Jupiter, they undergo a cycle of eclipse, occultation and transit. When a satellite passes into the shadow which Jupiter casts, it is no longer being illuminated by the Sun, so cannot be seen from Earth, even if it is not directly hidden by Jupiter. When this happens, the satellite is in **eclipse**. When it actually passes behind the disc of Jupiter, it is said to be **occulted**. Of course, the satellite also has to pass in front of Jupiter's disc, when it is said to be in **transit**. A large telescope will show up the black dot on the surface of Jupiter which is the shadow of the satellite.

6.2 The cycle of phenomena undergone by Jupiter's satellites as observed from the Earth
A. The satellite enters Jupiter's shadow and disappears in eclipse
B. The satellite reappears from occultation
C to D. The satellite is in transit across the disc of Jupiter
If the relative positions of the Sun, the Earth and Jupiter are different, the satellite may disappear in occultation and reappear from eclipse

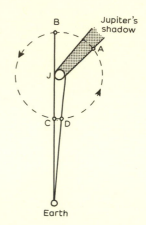

Io is an especially interesting satellite because it probably possesses an atmosphere. Also it exerts a noticeable influence on the strength of radio waves from Jupiter.

7 Saturn

Saturn is the second largest planet in the Solar System, with a diameter which is just under 10 times the Earth's. The physical conditions seem to be very similar to those of Jupiter, but Saturn is colder, as it is further from the Sun. Its rotation period is about 10 hours, just slightly longer than Jupiter's, so Saturn's poles are also considerably flattened. Unlike Jupiter, Saturn's atmospheric features are not at all permanent or conspicuous. Occasionally, observers have reported seeing white spots. Saturn is less dense than Jupiter, having a density which is even lower than that of water.

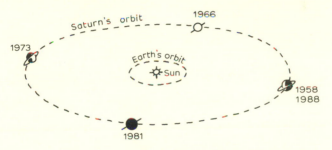

7.1 The orientation of Saturn's rings

Saturn's rings

The outstanding feature of Saturn is of course the system of rings. The rings, which circle around the planet's equator, are probably less than 15 km thick, and are quite transparent to starlight. They are definitely not solid, and are probably made of dust, and possibly larger rocks up to a few metres in size. There are two main rings, designated A and B, separated by the Cassini division. There are very faint inner sections (C and D) to the ring system as well. As the positions of Earth and Saturn change relative to each other, the ring system is presented to us at different angles. Every fifteen years the rings are seen edge on. At this time it is usually impossible to detect them at all. The brightness of Saturn in the sky is at its lowest then, too, as the rings contribute a considerable proportion of the sunlight which Saturn reflects.

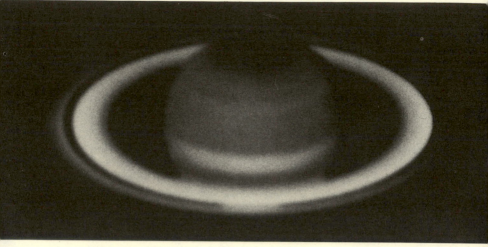

8. Saturn and ring system. (Hale Observatories)

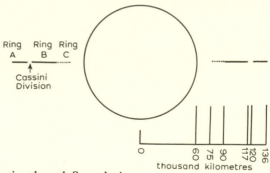

7.2 Cross-section through Saturn's rings

The satellites of Saturn

The largest satellite of Saturn, Titan, is about the size of Mercury and has the distinction of being one of only two planetary satellites known to possess an atmosphere. This is composed of hydrogen, nitrogen and methane. Titan is a prime candidate for exploration by spacecraft towards the end of the century.

There are nine other satellites, five of which are extremely tiny. The gravitational pull of the inner satellites is responsible for the Cassini division in the ring system.

8 *The discovery of the minor planets, Uranus, Neptune and Pluto*

Until the end of the eighteenth century, the only planets known to man, apart from the Earth, were the five known since antiquity which are easily seen with the naked eye. In 1781, Herschel discovered Uranus. Then on 1 January 1801, the Italian astronomer, Piazzi, discovered the first of the minor planets (or asteroids) which was later named Ceres. Since then, thousands of minor planets have been found, together with two more major planets.

The minor planets (asteroids)

Between 1801 and 1807 the four largest minor planets were discovered. They are, with their diameters, Ceres (770 km), Pallas (490 km), Juno (200 km) and Vesta (380 km). The fifth was not found until 1845, but then the numbers rapidly increased. The majority of minor planets are 100 km or less in diameter, and revolve around the Sun in almost circular orbits, close to the plane occupied by the major planets, and confined to a region between the orbits of Mars and Jupiter. However, there are a number of exceptions. Some, such as Eros, Adonis, Apollo and Hidalgo, have very eccentric elliptical orbits, and can occasionally pass close to Earth.

No one knows the origin of the minor planets. They are not likely to be the remains of a small planet that broke up since it has been estimated that the total mass of all the minor planets put together is at most one-hundredth of the mass of the Earth. Most likely they are wreckage left over from the formation of the planetary system.

The discovery of Uranus

Uranus was discovered accidentally by William Herschel on 13 March 1781. At first he thought it was a comet, but later observations showed it to be a new planet. Although it turned out that Uranus had actually been seen before, no one had recognized it as a planet. Planets show a visible disc in a telescope, whereas stars always appear as points of light. Comets are generally hazy and usually show a tail. Herschel's great skill as an observer enabled him to discern that the image of Uranus was larger than a star's. Herschel wanted to call the new plant Georgium Sidus in honour of his patron, George III, while others wanted to call it Herschel. Eventually the name Uranus was adopted by all, because in mythology Uranus was the father of Saturn, who was the father of Jupiter.

The discovery of Neptune

Uranus was observed carefully after its discovery, but soon it was found that it did not follow the predictions which had been made for its path in the sky. Furthermore, the fortuitous observations that had been made before Uranus was recognized, which were re-examined, did not seem to fit in either. In 1834, the Reverend T. J. Hussey wrote to G. B. Airy, who was professor at Cambridge, and pointed out the discrepancies between the predicted tables of the planet's position, and the actual observations. Airy thought that there might be another planet further away than Uranus whose gravitational pull was disturbing the orbit of Uranus. However, Airy did not think that the position of the planet could be calculated from the discrepancies. Nevertheless, the calculations were carried out by both John Adams, a young Cambridge mathematician, and Le Verrier in France. By this time, Airy had become Astronomer Royal and moved to Greenwich. He asked the new professor at Cambridge, James Challis, to search for the supposed planet. Initially it was thought that the planet was unlikely to be large enough to have a visible disc, so he started a long slow search, including observations of quite faint stars in the hope of detecting the planet by its motion among the stars. In those days, without the help of photography, it was very hard work mapping the sky. Le Verrier, however, later decided that the planet might have a visible disc. He wrote to Dr Galle at the observatory in Berlin, who, with the aid of newly prepared charts, found the new planet immediately, on 25 September 1846, near the place that Le Verrier had predicted. It turned out that Challis had observed Neptune in August 1846, but had failed to recognize it as a planet, or to analyse his results immediately.

The discovery of Pluto

As the paths of Neptune and Uranus were followed, slight discrepancies in *both* their orbits were discovered, which could not be explained by the gravitational pulls of the known planets. So naturally, the speculation was that yet another planet, the so-called 'planet X' might be responsible. The search for planet X was started in 1906 by Percival Lowell at the observatory which now bears his name at Flagstaff, Arizona. Lowell died in 1916 without finding it. The only way to find this planet was to look for the slight motion against the background of stars which would show up between photographs taken some time apart. It was necessary to compare a large number of photographs, so an instrument called a blink comparator was used. In this device, the two photographs being compared are illuminated alternately, very rapidly. If anything is in a different position on one of the photographs, the operator sees the object apparently jumping back and forth.

In 1919, E. C. Pickering made further calculations and started an observing program at the Mount Wilson Observatory, but he failed to notice Pluto, though it turned out later that he had photographed it. Pluto was not, in fact, in the most likely place according to his calculations, so he missed it. In 1929, C. Tombaugh, at the Lowell Observatory, started yet another systematic search, taking three photographs of each part of the sky in which he thought planet X might lie. In 1930 his patience was rewarded. He discovered the planet, which was given the name Pluto, in the constellation of Gemini.

The physical nature of the outer planets

Comparatively little is known about the three outer planets because of their vast distance. Uranus and Neptune are thought to be similar in constitution to Jupiter and Saturn, but colder of course. Both have atmospheres which are rich in methane. Pluto is probably more like a deep-frozen version of the Earth or Mars. The most remarkable feature about Uranus is that its axis of rotation is inclined at 98° to the plane of the Solar System, so the planet appears to be rotating on its side. Uranus has five satellites and Neptune two. Pluto is so small and distant that a disc is only resolvable with the Hale Observatories' 5-metre telescope, so it is unlikely that much will be found out about Pluto in the near future.

In 1977 a rare event took place, the occultation of a star by Uranus. Uranus moves so slowly through the stars that it is very unusual for an occultation to occur. This one was observed by several teams of astronomers who all agreed on an unexpected discovery. Uranus, like Saturn, is encircled by a ring system. Unlike Saturn's rings, though, those of Uranus are narrow, separated by wide gaps. Five rings were found and they were named after the first five letters of the Greek alphabet, Alpha, Beta, Gamma, Delta and Epsilon. It is not possible to detect the rings by ordinary observation, even with the world's largest telescopes.

9 The Earth as a planet

From space, the Earth appears as a blue disc with the characteristic swirling patterns of white cloud. The blue colour arises from the scattering of sunlight by the atmosphere. From moderate distances the land masses and oceans can be distinguished. Earth is unlike any of the other planets in having the major depressions of its surface filled by liquid water, which covers 80 per cent of the planet. In general, though, the depth of trenches (10 km maximum) and height of mountains (8 km maximum) are insignificant compared with the over-all size of the Earth (12,600 km diameter).

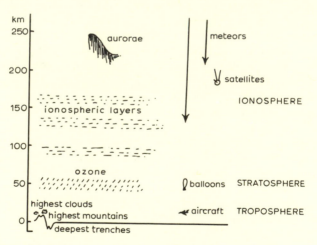

9.1 The structure of the Earth's atmosphere and the heights of various objects and phenomena

The atmosphere

The Earth's atmosphere is composed of 78 per cent nitrogen, 21 per cent oxygen, 0·9 per cent argon with carbon dioxide and water vapour in variable amounts. In the outermost parts there is a little hydrogen and helium. The atmosphere protects the Earth from the constant barrage of ultraviolet light and X-rays which arrive from the Sun and from the distant Universe. These radiations are absorbed in the upper atmosphere, where they **ionize** the air molecules. In this process, electrons are stripped from the molecules, thus leaving electrically charged particles which form a protective layer called the ionosphere. Below the ionosphere is a region containing ozone (oxygen molecules containing three atoms instead of the more usual two) which is also created by the action of ultraviolet radiation. The atmosphere also protects the Earth from the devastating impact of meteors. They are heated up by friction as they enter the atmosphere at tremendous speeds, up to 60 km/s, and burn to dust before reaching the ground. The absorption of electromagnetic radiation by the atmosphere limits astronomical observation outside the visible part of the spectrum. Only limited parts of the infrared and radio regions are transmitted, so X-ray and ultraviolet observations have to be made on rockets, balloons or satellites.

The interior structure

The interior structure has been determined largely by the study of earthquakes and the way in which earthquake waves travel (seismology). Such studies have revealed that there are several major discontinuities in the interior. The mean density of the Earth, 5,500 kg/m³, is larger than the density of the surface rocks. This further confirms the seismic evidence which suggests that the inner layers of the Earth are more dense than the outer ones. The core of the Earth probably consists of iron and nickel. Electric currents flowing in a metal core would account for the Earth's magnetic field. The core is thought to be at least partially liquid, if not entirely so, and the interior is still very hot from the time of the Earth's formation. Radioactivity is thought to be a continuing source of the heat energy.

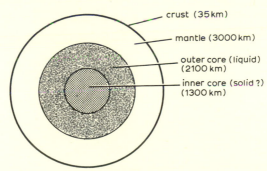

crust (35 km)

mantle (3000 km)

outer core (liquid) (2100 km)

inner core (solid ?) (1300 km)

9.2 Interior structure of the Earth

The magnetic field

The magnetic field of the Earth is like that of a gigantic bar magnet (what is known as a 'dipole' field). This field probably originates from electric currents flowing in the metallic interior of the Earth. At the moment, the Earth's magnetic axis is inclined at about 12° to its axis of rotation, but it does move about in an irregular way. There is evidence in rock formations that during the geological past, the Earth's magnetic field has completely reversed a great many times.

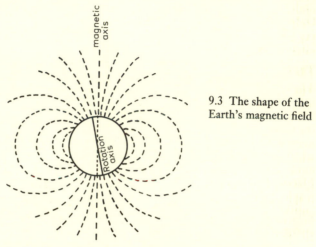

9.3 The shape of the Earth's magnetic field

When electrically charged particles encounter a magnetic field, they travel in helical paths around the lines of magnetic field. Charged particles which stream out from the Sun are always arriving near the Earth, especially at times of increased solar activity. Many of these particles (chiefly electrons), become trapped by the Earth's magnetic field in particular regions which surround the Earth like concentric rings. These are called the **Van Allen belts**. It is these trapped particles which are responsible for the **aurorae**.

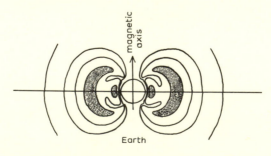

9.4 Cross-section through the Van Allen belts

The cross-sections of the Van Allen belts are shaped like crescents. The 'horns' of the crescent nearest Earth point towards latitude 73° N and S. Near the horns, the particles can be guided down to the Earth along the lines of the magnetic field. They are moving very rapidly, and collide with atoms in the upper atmosphere thus causing the atoms to send out light. This explains why the regions where most aurorae are observed are near to the arctic and antarctic circles, and why more are seen when solar activity increases.

10 *Comets*

Most comets can only be seen with a telescope, but occasionally one becomes bright enough to be visible to the naked eye. Each comet has its own individual appearance which may change in the course of the few weeks during which it is visible. The general structure of all comets, however, seems to be more or less the same. A comet consists of a bright head, surrounded by a more diffuse coma, and a tail which usually develops as the comet approaches the Sun, streaming in the direction away from the Sun, often over a vast distance.

10.1 The parts of a comet

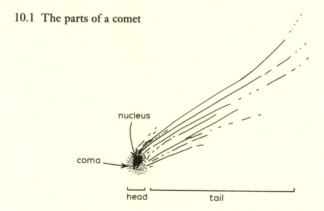

The brightness of a comet increases rapidly as it approaches perihelion. Much of the light is reflected sunlight, but there may be a proportion of light generated within the comet itself, in response to solar heating. At any given time the observed brightness of a comet depends both on its distance from the Sun, which determines the comet's luminosity, and its distance from Earth, which affects the apparent magnitude. Occasionally a comet may develop a huge tail, perhaps 100 million km long, as it nears the Sun, and this will greatly increase the apparent magnitude.

9. Comet Ikeya-Seki, 1965, photographed from Mount Wilson Observatory, California. The lights are the city of Los Angeles and the surrounding area. (Courtesy W. Liller)

Origin of comets

The origin of comets is unknown and has been the subject of intense speculation. They are the most erratic members of the solar System, and have been observed since antiquity. In ancient times they were thought to be atmospheric phenomena. Tycho Brahé demonstrated, in the sixteenth century, that this idea was incorrect when he managed to show that comets were further away than the Moon. It was Edmund Halley who calculated the motion of comets and showed that they were members of the Solar System, moving in accordance with Newton's law of gravitation.

One theory for the origin of comets suggests that there may be a vast shell of comets at the farthest reaches of the Solar System. The shell is said to be half-way to the nearest stars, beyond the limits of detectability. When a comet gets disturbed from the shell, it may head towards the Sun.

Another idea is that comets are manufactured in our Solar System as the Sun journeys around our Galaxy. Every few million years the Sun passes through dense clouds of dust which exist in space. The action of the Sun's gravitational field may cause new comets to condense from the dust cloud.

Composition of comets

Comets appear to be made of gas, dust and various ices, in proportions that vary from object to object. The mass of comets is very low indeed, and any substantial concentration of matter, such as an icy or rocky nucleus, cannot be more than a few kilometres across, which is too small for direct observation. As a comet approaches the Sun the icy substances are evaporated and dust activity increases.

The composition can be determined, at least partly, by spectroscopic analysis of the light from a comet. Simple molecules containing carbon, nitrogen, oxygen and hydrogen seem to be the most common (e.g. C_2, CH, NH, OH, CO). The tail develops spectacularly when the comet is close to the Sun. It is thought that heating by the Sun and the effect of the streams of particles from the Sun (the 'solar wind') are responsible for the tail, which is composed of gas or dust or both.

Each time a periodic comet goes round the Sun it loses material, perhaps as much as one per cent on each passage. This substantial wastage means that comets become dimmer if they pass the Sun many times, and this has certainly happened for Encke's Comet and Halley's Comet. Since comets are short-lived astronomically, and since they sometimes disintegrate, the comet population must be being continuously re-plenished by new comets.

Orbits

The majority of comets appear to be in parabolic orbits with the Sun at the focus. Strictly speaking, a parabola is an open curve, but if the orbits of the comets are really very elongated ellipses, we would interpret as parabolas the small sections of the orbits that can be observed. Whatever the true shape of their orbits, comets really belong to the Solar System.

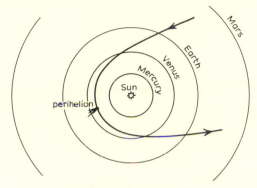

10.2 Parabolic comet orbit

A comparatively small number of comets are in true elliptical orbits, with periods ranging from 3 to 250 years, and these are called periodic comets. Of these, a considerable number have periods between 7 and 15 years. These have been captured into the inner Solar System by the strong gravitational pull of Jupiter. It is thought that all periodic comets have at some time been trapped into their orbits accidentally by passing close to a planet.

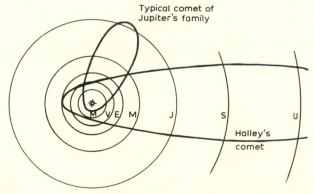

10.3 Elliptical comet orbits

The orbits of the major planets all lie very nearly in one plane. Cometary orbits, though, may be inclined at any angle to the plane of the Solar System. The angle between the planes of a comet's orbit and the Earth's orbit is called the inclination. The point of closest approach to the Sun is termed **perihelion**. For some comets, called 'sun-grazers', the distance from the Sun at perihelion is so small that the comet passes through the outer part of the Sun. With such a comet there is always the danger that it will totally disintegrate and fall into the Sun.

Naming of comets

Comets are named after their discoverers. Perhaps that is why so many amateur astronomers spend many hours patiently sweeping the sky with binoculars in the hope of finding one. Periodic comets have the prefix 'P/' before their names. Each comet is also ascribed a provisional letter to indicate the order of discovery (or recovery in the case of a known periodic comet). Later, each comet is given a roman numeral instead of its letter. These numbers are in order of perihelion passage during the year.
e.g. Comet Alcock 1963b became 1963 III
     Periodic comet P/Halley 1909c became 1910 II

11 *Meteors*

Meteors are small rocks or dust particles in the Solar System. The impact of meteors is thought to be responsible for many of the craters on the Moon, Mercury and Mars. Twenty million meteors hit the Earth's atmosphere daily, but few ever reach the ground. They enter the atmosphere travelling so fast that friction between the meteor and the air causes the meteor to burn up, and most do not get below a height of 100 km. The burning meteor can often be seen as a bright streak across the sky lasting a few seconds, hence the common name of 'shooting star'. Sometimes the streak is coloured reddish or bluish.

Meteorites

Meteors that do not burn up completely and therefore reach the ground are called **meteorites**. Meteorites are of two main types. Iron ones are composed almost entirely of iron and nickel; they are much denser than ordinary rocks and are easily recognized. Stony meteorites are thought to be much more numerous, but fewer are actually found as they closely resemble ordinary rocks. There are a few large craters on the Earth that have been created by the impact of very large meteorites, for example the famous Barringer Crater in Arizona, which has a diameter of 1·6 km. The largest known meteorite fell in South Africa in prehistoric times and weighs 60 tonnes.

Meteor showers

The Earth sometimes passes through a stream of interplanetary particles, a phenomenon which may cause a **meteor shower**. In a shower, a large number of meteors are seen in a short space of time. Most showers last a day or two and at best there are one or two meteors every few minutes. As all the meteors are coming from one direction in space, as viewed from the Earth, the meteor tracks all seem to diverge from one point in the sky. This point is called the **radiant** of the shower. Meteor showers are often associated with the debris left strung out in the orbit of a comet, and so the showers recur each year at the same time, with the same radiant, when the Earth intersects the orbit of the comet. There are, for example, two regular showers associated with Halley's Comet. They are the Aquarids and the

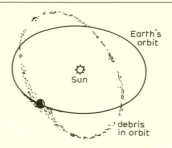

11.1 How meteor showers occur

Orionids. The Perseids, which appear each year in late July and early August, are associated with Comet 1862 III. Shower names are derived from the constellations in which the radiants lie.

Apart from showers, meteors can be seen coming randomly from all directions, all the year round. If you stand outside for half an hour on a dark and clear night, you will almost certainly see one. These random meteors are called **sporadic meteors**. The most comfortable way to observe meteors is to lie inside a sleeping-bag on a deck chair.

12 *The Earth's motion in space*

Each year the Earth completes one revolution of its continuous journey round the Sun. We recognize the passage of a year by the cycle of seasons, or by the changing pattern of stars in the evening sky or by counting off the days on a calendar. One **solar day** is the time it takes for the Earth to rotate once on its axis *relative to the Sun*. There are 365·24 days in a year. In the space of each day the Earth moves on in its orbit, so that the apparent position of the Sun relative to the fixed stars is changing continuously. The Earth's spin is in the same direction as its orbital motion. It follows that it

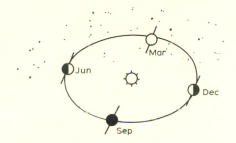

12.1 During the course of a year the night side of the Earth looks out on different star patterns

takes *longer* for our planet to turn so that the same point on Earth is once again facing the Sun, than it does to turn so that people at a certain place can see the same *stars* once again. Another way of putting it is that after 24 hours, the night side of the Earth is pointing in a slightly different direction in space compared to the previous night, so observers get a slightly different view of the sky on successive nights. In fact, it takes only 23h 56m for the Earth to turn on its axis *relative to the stars*, and this length of time is called a **sidereal day**. The Earth's orbit round the Sun carries the Earth round on its axis relative to the stars, an extra time; so there are 366·26 sidereal days in a year.

The seasons

Seasons occur because the plane of the Earth's equator is inclined at an angle of about 23½° to the plane of its orbit. Except for very long-term

N
Sun's
rays
S
Summer
in north

N
Sun's
rays
S
Summer
in south

12.2 The tilt of the Earth's rotation axis at 23° to the
plane of its orbit causes the seasons

changes called **precession**, the Earth's axis always points to the same
direction in space. Because of the Earth's motion round the Sun, in the
northern hemisphere's summer the north pole is tilted towards the Sun,
whereas in winter the pole is tilted away from the Sun. The opposite is true
in the southern hemisphere. The seasonal changes are familiar to
everyone. In summer the Sun gets higher in the sky and the days are
longer. In fact, in the arctic and antarctic circles the Sun stays above the
horizon for the whole 24 hours. In winter, however, these regions
experience 24 hours of night. It is not only the length of the day which
affects the warmth of the weather. When the Sun's altitude is low, the
radiation has to penetrate a greater thickness of atmosphere, and it strikes
the ground at a shallow angle so that it is spread over a greater area, which
leads to less intense heating.

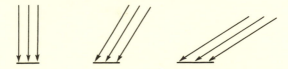

12.3 Dilution of the Sun's rays when the Sun's altitude is low

Precession

The direction in space towards which the Earth's rotation axis points is not
constant. The Earth is not a perfect sphere, but is rather flattened at its
poles. Because of the combined attractions of the Sun and Moon on
Earth's equatorial bulge, the rotation axis slowly swivels around a circle
47° in diameter, in a period of 25,800 years. The axis sweeps out a cone in
space with a semi-angle of $23\frac{1}{2}°$. This phenomenon is called **precession**.
It causes a gradual change in the seasons known as the precession of the
equinoxes. In about 13,000 years' time summer will occur in the months
that are winter now. Also there is a change in the direction of the north

celestial pole, so that Polaris (the Pole Star) will not always be near the north pole. In the time of ancient Egypt, for example, Alpha Draconis (Thuban) was the pole star, and in about 12,000 years' time Vega will be close to the pole.

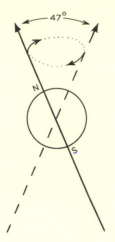

12.4 Precession. The Earth's rotation axis sweeps out a cone in space of diameter 47°, in a period of 25,800 years

13 *The Night Sky*

The celestial sphere

The stars we see in the night sky are scattered through space, all at different distances from the Earth. However, they are all sufficiently far away that for the purpose of mapping the sky we can imagine that they are stuck on the inner surface of a distant transparent globe. In fact, before the invention of the telescope it was universally assumed that the stars were fixed on a translucent crystal shell. It is this imaginary globe that is called the **celestial sphere**.

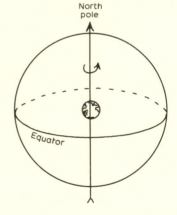

13.1 The imaginary celestial sphere

Sky coordinates

Any place on the Earth can be conveniently located on a map by means of its latitude and longitude. Similarly, any object in the sky can be found if its coordinates on the celestial sphere are specified. The celestial equivalents of latitude and longitude are called **declination** and **right ascension**. The celestial globe has its own north and south poles and equator, and these are defined as follows:

Imagine extending the Earth's axis so that it points right out into space, towards the fairly bright star called Polaris. The Earth's axis points to the

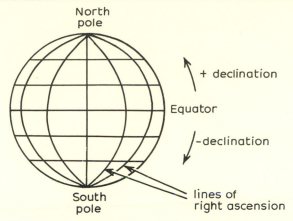

13.2 Right ascension and declination on the celestial sphere

north celestial pole. Now imagine the Earth's equator projected out on to the celestial globe; the projected circle defines the **celestial equator**. Just like latitude, declination is measured in degrees north and south of the celestial equator, except that 'north' and 'south' are not used. Instead northern declinations are denoted by a + sign and southern ones by a − sign. Just like lines of longitude, lines of right ascension stretch from pole to pole. However, for reasons which will become clear later, right ascension is measured not in degrees, but in hours and minutes of *time*. The 360° are divided into 24 hours, so each hour of right ascension is equivalent to 15°. Just as zero of longitude is measured from the Greenwich meridian, astronomers have had to choose the zero point from which to measure right ascension. The zero of right ascension is defined as the point where the Sun, in its annual motion through our skies, crosses the celestial equator moving from south to north. The Sun does this on or about 21 March according to our calendar. Because this day is one of the two equinoxes the point 0 hours on the celestial equator is often called the **vernal equinox**. Its other name is **'the first point of Aries'** because when it was first chosen it lay in the constellation Aries, although it no longer does so on account of precession. This point in the sky has its own special symbol, ♈, the old symbol for the zodiac sign Aries, the ram.

The observer's hemisphere

An observer on the Earth can see only half of the celestial sphere at any one time. Even over a period of time he is never able to see all of the heavens unless he is on the equator. The view of the sky depends on the *latitude* from which you observe. The altitude of the north pole is equal to the observer's latitude on the Earth. For example, at the north pole of the Earth, the north pole of the sky is directly overhead. The celestial equator stretches across the sky from east to west. The imaginary semi-circle

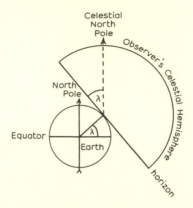

13.3 For an observer at latitude λ on the Earth, the celestial north pole is $\lambda°$ above his horizon

through the sky over the observer's head from north to south is called the **observer's meridian**. The point directly over his head is the **zenith**. That directly below his feet, in the half of the sky he cannot see, is the **nadir**. Zenith and nadir, like most star names, are Arabic words, a legacy from the great Arabic work in astronomy during the Dark Ages in Europe.

The effect of the Earth's rotation

Relative to the stars, the Earth rotates on its axis once in 23h 56m. To an observer on the Earth, it seems that the whole sky rotates about the celestial poles in 23h 56m, which are the only points that seem to be still. For an observer at northern latitudes, there are stars near the north pole that sweep out circles around the pole, and they never set. These stars are **circumpolar**. It can be shown easily with a diagram that only stars whose declinations are larger than $(90 - \lambda)°$ are circumpolar at a place of latitude λ. All other stars rise in the east and set in the west. When a star crosses the observer's meridian, it is said to **transit**, or to **culminate**. Circumpolar stars cross the meridian *twice* a day, and have both upper and lower transits. When it is on the meridian at upper transit, a star is at the greatest altitude it ever reaches, and so is in its best position for observation.

Altitude and azimuth

Right ascension and declination are not the only coordinates that can be used to specify a star's position. They have the fundamental advantage of remaining practically constant during the course of observing a particular object. However, at any particular time, it may be useful to know a star's

10. Star trails around the south celestial pole. The exposure was about 8 hours during which time the stars swept out arcs of about 120° as the Earth rotates on its axis. (Anglo–Australian Observatory)

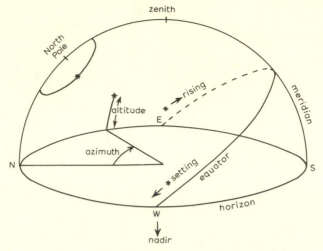

13.4 The observer's hemisphere (northern latitudes)

altitude and **azimuth**. The altitude is the angular distance above the horizon. The azimuth is the angular distance round the horizon from north (through east-south-west). These coordinates may be useful, for example, when assessing whether a star is visible from a particular place at some time, but of course they are changing continuously throughout the observing session.

Constellations and star names

We think of constellations in a general way as being groups of bright stars which seem to make conspicuous patterns in the sky. Many have the names of legendary characters or animals. In a stricter sense, a constellation is a clearly defined area of sky, with boundaries drawn up by international agreement. The whole celestial sphere is divided up into 88 constellation areas. In astronomy, the Latin constellation names are used. Quite a large number of bright stars also have their own proper names, although only a limited number of these names are now in common use. These proper names are mostly Arabic in origin. The most commonly used way of designating the brighter stars is by the Greek alphabet, and the constellation names. The brightest star in each constellation is designated 'α' followed by the genitive case of the Latin name meaning 'of the constellation . . .'. For example, the brightest star in the constellation Leo is α Leonis. Often, though, a standard 3-letter abbreviation is used for the constellation (e.g. UMa for Ursa Major, the Great Bear). The next brightest stars in the constellation are designated β, γ etc. in a similar way, until there are no more Greek letters left. Fainter stars are often referred to by their number in a catalogue published in 1725 by Flamsteed, who was Astronomer Royal (e.g. 32 Virginis). Many variable stars have names made up from one or two capital letters between R and Z (e.g. RR Lyrae). With the advent of astronomical photography it became possible to produce catalogues listing hundreds of thousands of faint stars, and these are invaluable to professional astronomers.

14 *The paths of the Sun, Moon and planets*

The Sun, Moon, and planets are all in relative motion, and they are also very close to each other, compared with the immense distances to even the nearest stars. As a result of their proximity the Sun, Moon and planets sweep out their paths across a background of apparently fixed stars.

The path of the Sun

The equator is inclined at $23\frac{1}{2}°$ to the plane of the Earth's orbit around the Sun, so that as viewed from the Earth, the Sun's path through the sky is a circle inclined at $23\frac{1}{2}°$ to the celestial equator. The path of the Sun is called

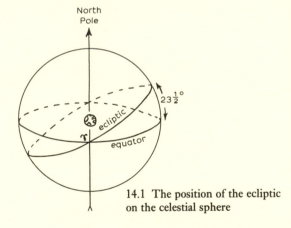

14.1 The position of the ecliptic on the celestial sphere

the **ecliptic**. This name arose because it was noted that eclipses could only occur when the Moon crossed this circle in the sky. It takes the Sun one year to sweep out the whole of its 360° path. Of course, we cannot see stars during the day because of the brightness of the Sun, but that does not mean that they are not there! The ring of constellations through which the ecliptic passes is called the **zodiac**, and the 'birth signs' quoted by astrologers are based on the constellations in which the Sun can be found during different times of the year. During a total eclipse the stars in the vicinity of the Sun can be seen.

There is a seasonal change in the Sun's declination which gives rise to the change in day length, and altitude of the Sun. On 21 March and 21 September, the Sun is on the equator, at declination 0°. These dates are called the vernal and autumnal **equinoxes**, because the position of the Sun means that day and night are of almost equal length everywhere. On these dates, and these dates alone, the Sun rises due east and sets due west everywhere. At the equator, however, the Sun always rises due east and sets due west. On 21 June the Sun reaches its maximum declination of $+23\frac{1}{2}°$ and on 21 December its minimum declination of $-23\frac{1}{2}°$. These two dates are the solstices.

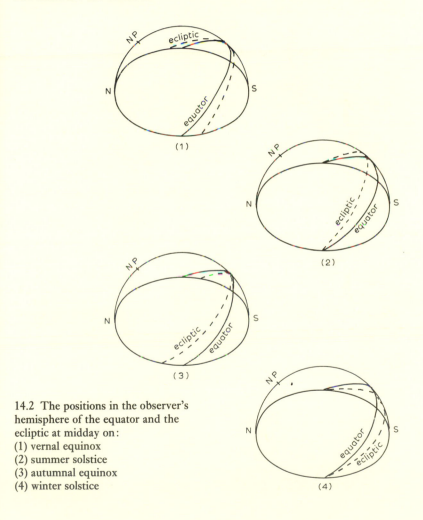

14.2 The positions in the observer's hemisphere of the equator and the ecliptic at midday on:
(1) vernal equinox
(2) summer solstice
(3) autumnal equinox
(4) winter solstice

Now it is easier to understand the definition of ♈, the zero of right ascension. Strictly, the vernal equinox is the intersection of the equator and the ecliptic, and 0 hours is the right ascension of this intersection. Unfortunately, the picture is complicated by the effects of precession. The gradual change in direction of the Earth's axis causes the ecliptic to 'slide round' the equator, so that it makes one complete circle in 25,800 years, and is always inclined at about $23\frac{1}{2}°$ to the equator. This means that the zero of right ascension is gradually and continuously moving so that the right ascensions and declinations of stars are steadily changing. The rate of precession is sufficiently great for its effects to be noticeable in only a day in accurate positional astronomy. Since the zero of right ascension was first defined, it has moved out of the constellation Aries, although it is still called 'the first point of Aries'! Another effect of precession has been that the 12 classical astrological birth signs no longer correspond to the actual location of the Sun in the zodiac. In fact, the Sun's path now passes through 13 constellations. The new one is Ophiuchus.

The path of the Moon

The orbit of the Moon around the Earth is inclined at 5° to the ecliptic. This means that the Moon's declination ranges between $+28°$ and $-28°$, but in fact the details of the Moon's motion are complicated and one whole cycle of motion is only completed in 18·61 years. Within this longer cycle, the Moon completes one circle against the background of stars each **sidereal month**, which is 27d 7h 43m. The interval between successive occurrences of the same *phase* (the synodic month) is about 2 days longer. (See Section 2.)

The paths of the planets

The orbits of all the planets, except Pluto, lie very nearly in one plane, so that as viewed from the Earth, their paths never stray far from the ecliptic. The rate of progress through the stars depends on the planet's orbital period. The inner planets, Mercury and Venus, move very rapidly, and the Earth is orbiting as well so that their paths appear quite complex. The rate of travel of the planets against the stars decreases sharply with their distance from the Sun. The planets beyond Mars may stay in the same constellation for years.

An interesting feature of the paths of the outer planets is their **retrograde motion**. Sometimes a planet will appear to come to a halt, and then travel backwards for a while, before continuing in the normal direction, thus tracing out a loop on the sky. This action results from the *relative* motion of the Earth and planet.

A superior planet (that is, one whose orbit is outside the Earth's) is said

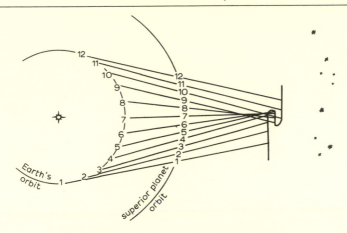

14.3 Retrograde motion of a superior planet

to be at **opposition** when it forms a line with the Earth and Sun, on the same side of the Sun as the Earth. As Mars' orbit is quite elliptical, some oppositions are much closer to the Earth than others. Oppositions of Mars occur about every two years. The time between successive oppositions of a planet is called its **synodic period.** For the more distant planets, this period is just over one year, because they travel only a small portion of their orbits in the time it takes the Earth to complete one journey round the Sun.

15 *Keeping time*

Our chief measurement of time is by the apparent motion of the Sun through the sky. We divide the day into 24 equal hours, but the Sun's progress is not at a steady rate. This is because the Sun moves along the ecliptic, and not along the celestial equator, and also because the Earth's orbit around the Sun is elliptical, not circular. So for the sake of convenience time is kept by an imaginary **mean Sun**, which is moving at a regular speed and keeping **mean solar time**. Time judged by the position of the *real* Sun is called **apparent solar time**. The difference between the apparent and mean solar times is called the **equation of time**; it ranges between about plus and minus 12 minutes during the year. This means that time as measured by a sundial has to be corrected by the equation of time to give mean time.

Midday is taken as the time when the mean Sun crosses the observer's meridian. This means, of course, that the time is different for people at different longitudes on the Earth. As a 360° turn of the Earth takes exactly 24 hours, a difference in longitude of 15° between two observers makes a time difference of 1 hour. It would be very inconvenient if everyone went by their own personal local time, as determined by the passage of the Sun across their own meridian, so the Earth is divided into 15° time zones. Everyone within a zone takes the same **standard time**. The boundaries between the time zones are not always meridians but take account of the location of towns and national frontiers.

Even use of standard time is confusing for astronomers, because there is frequently need to compare the times of observations made in different parts of the world. Instead astronomers all use the time at longitude 0°. This is called **Universal Time**, and is effectively what was formerly called Greenwich Mean Time.

Sidereal time

When astronomers are planning their observations, they are primarily interested in the progress of the stars through the sky. Relative to the stars the Earth rotates once every 23h 56m. Time kept by successive meridian transits of the stars is called **sidereal time**. A clock keeping sidereal time has to run slightly faster than one keeping solar time, gaining about 4 minutes per day. Now it is clear why right ascension is measured in hours and minutes. The local sidereal time is *defined* as the right ascension of the local meridian. Solar and sidereal times are only in step at the autumn equinox, when 0h of right ascension coincides with the meridian at midnight.

The sidereal time changes according to the observer's longitude in exactly the same way that solar time does.

Hour angle

The difference between the sidereal time (i.e. the R.A. of the meridian) and the R.A. of an object is called the object's **hour angle**. The origin of the name is obvious because it gives the time which has elapsed since the object was on the meridian.

Hints on solving problems concerning the celestial sphere and time

These problems are not as difficult as they appear at first provided that a systematic approach is adopted.

The following suggestions should be found useful:

(1) If it is at all relevant, draw a diagram of the appropriate celestial hemisphere, or the whole sphere, even if you are not asked to do so specifically.

(2) Mark in the north pole, remembering that its altitude is governed only by the observer's latitude. Then mark in the equator, the meridian and anything else that seems relevant.

(3) Where time is concerned, remember that for every degree of longitude west, the time (either solar or sidereal) is 4 minutes later, and for every degree of longitude east it is 4 minutes earlier.

16 *Kepler and Newton, and the Laws of Gravity*

The true nature of the orbits of the planets was first deduced by Johannes Kepler (1571–1630). He used the observations made by his master, the great Danish astronomer Tycho Brahé (1546–1601). Kepler drew up three empirical relations which describe the motion of the planets. They are:

First Law : **The orbit of each planet is an ellipse with the Sun at one of its foci.**

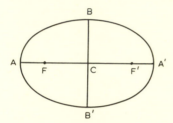

16.1 The geometry of the ellipse AC = A′C = a (semi-major axis) BC = B′C = b (semi-minor axis). F and F′ are the foci. FC = F′C = a × e where e is the *eccentricity* of the ellipse. The larger the value of e, the more oblate the ellipse. e is always less than 1, and is 0 for a circle

Second Law : **The radius vector** (that is the imaginary line which joins the Sun to the planet) **sweeps out equal areas in equal times.**

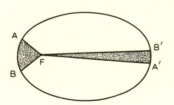

16.2 Kepler's second law of planetary motion
The areas of the two sectors AFB and A′FB′ are equal. As the planet's radius vector sweeps out equal areas in equal times, the planet must move faster when it is close to the Sun

Third Law : **The squares of the sidereal periods of the planets are proportional to the cubes of their mean distances from the Sun.** Expressed mathematically:

$$P^2 \propto a^3$$

or P^2/a^3 is a constant for all the planets.

Newton and gravitation

The theoretical basis for the relationships which Kepler had found empirically was only established later when Isaac Newton (1642–1727) formulated the Law of Universal Gravitation. According to this, all masses experience a mutual force of attraction, whose size, F, is given by the relationship

$$F = G\frac{m_1 m_2}{r^2}$$

m_1 and m_2 are the two masses which are separated by a distance r, and G is the universal gravitational constant.

It is the force of gravity which keeps the planets in their orbits around the Sun and prevents things on the surface of the Earth from floating away into space.

Orbital motion

Newton was also responsible for formulating the laws of motion which are still universally valid. According to the First Law of Motion, **everything continues in its state of rest or uniform motion in a straight line unless it is acted on by a force**.

If an object is going to move in a circle, or an ellipse, rather than a straight line, it can only be compelled to do so by the continual action of a force directed inwards. This force is called the **centripetal force**. In the case of a planet orbiting round the Sun, this force is provided by the gravitational attraction of the Sun for the planet.

Meanwhile, an object which is rotating, for example a person standing on the spinning Earth, experiences an *outward* force, equal in magnitude to the centripetal force. This force experienced by a rotating object is called the **centrifugal force**.

The centripetal force F, on an object of mass m, moving with a velocity v in a circle of radius r is given by the relation:

$$F = \frac{mv^2}{r}$$

By use of this relation and the Law of Gravitation, it is not difficult to show mathematically the truth of Kepler's Third Law, in the case of a circular orbit.

16.3 Circular orbit

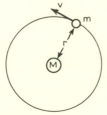

Kepler's Laws follow automatically from the more basic law of gravity. The mathematical proof is more complicated for the general case of an elliptical orbit, but the result turns out to be just the same as for a circular orbit.

The gravitational field

The region around a mass in which a second mass would feel a gravitational force is called a **gravitational field**. The strength of the gravitational field of an object depends on how big its mass is, and, of course, gets less further from the mass. The gravitational field strength, g, is defined as the force which a mass of one unit would feel at that particular place,

thus,
$$g = G\frac{M}{r^2}$$

at a distance r from the centre of a mass M. The more common name for g is 'the acceleration due to gravity' because, in the absence of other forces such as air resistance all objects falling towards the mass have this same acceleration. According to legend, Galileo demonstrated the truth of this statement by dropping objects of different mass from the leaning tower of Pisa. The Apollo 15 astronauts repeated the experiment on the Moon, and showed that a hammer and a falcon's feather fell with the same accelerations.

According to Newton's Second Law of Motion, when an object of mass m experiences a force F, it acquires an acceleration, a, such that $F = m \times a$. If we apply this to the gravitational force, we can derive an expression for g:

$$G\frac{mM}{r^2} = m \times a$$

so
$$a = \frac{GM}{r^2} = g$$

For the Earth's surface g has an average value of $9 \cdot 8 \text{ m/s}^2$ although it varies slightly from place to place because the Earth is not of uniform shape nor of regular distribution of mass. Furthermore, as the Earth is rotating, the centrifugal force slightly lowers the effective acceleration due to gravity. The centrifugal force is greatest at the equator, reducing to zero at the poles. On the Moon, whose mass and radius are both less than the Earth's, g is about one-sixth its value on Earth.

Escape velocity

An object thrown upwards usually falls back to the ground. The faster the object is thrown, the higher it gets. There is a certain velocity of projection, which if exceeded, results in the object *not* returning, but escaping from the Earth's gravitational pull completely. This is the **escape velocity**.

The escape velocity from a planet of mass M and radius r is given by the equation

$$v_{escape} = \sqrt{\frac{2GM}{r}}$$

17 *The Sun*

The Sun is the only star whose surface can be examined in any detail. All other stars are so far away that they only appear as points of light, even in the largest telescopes. Nevertheless, the comparison between the light received from the stars and from the Sun leads to the conclusion that the Sun is a typical star.

The photosphere

The visible, sharp-edged disc of the Sun is called the **photosphere**. Photographs of the photosphere show a general decrease in brightness from the centre to the edge. This is known as **limb darkening**. The term 'limb' means the edge of the disc. The phenomenon arises because the outer layers of the Sun are not completely transparent. At the centre of the visible solar disc the line of sight penetrates to hotter, brighter layers than at the edge, where the line passes trangentially through cooler, dimmer layers.

Under magnification, the photosphere shows a mottled appearance. This effect is known as **granulation**.

Solar activity

The photosphere is frequently marked with dark spots; other transient phenomena occur which are associated with the appearance of spots. At this time too, streams of particles and energetic radiation may leave the Sun and cause disturbances in the Earth's atmosphere, which upset radio communications. These happenings are collectively known as **solar activity**.

Sunspots

The dark spots that appear on the Sun are areas which, at 4,000°K, are some 2,000 degrees cooler than the surrounding photosphere. They only look dark by comparison with the brilliant photosphere. They occur typically in groups starting as tiny specks which develop in size and structure, until, after several weeks, they gradually disappear. A large sunspot may be many times the size of the Earth. Even a small telescope, used to project an image of the Sun, will show that the spots consist of a

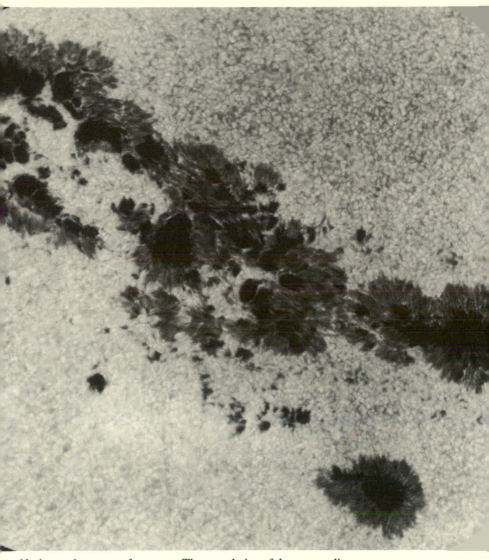

11. A complex group of sunspots. The granulation of the surrounding photosphere can also be seen. (Sacramento Peak Observatory)

very dark central 'umbra' surrounded by a brighter 'penumbra' which is marked with radial striations. Sunspots are occasionally so large that they extend across one-third of the Sun and can be seen by the unaided eye at sunset. They have been known since antiquity.

Plages

In the vicinity of sunspots there is often a region of photosphere which is hotter and brighter than general. Such a bright area is called a **plage**. Plages usually appear before their associated spots, and disappear later than them. Sometimes they appear without the occurrence of spots.

Prominences

These are great tongues of solar material which leap outwards from the surface of the Sun. They occur at the active plage regions. Seen against the background of the photosphere they appear as dark filaments, but at a total eclipse, prominences may be seen projecting from the limb. The largest prominences extend for more than a million kilometres.

Solar flares

Flares are sudden bright flashes which occur in active regions, lasting from a few seconds to a few minutes. They can be of great violence and intensity. During a flare, not only is there an increase in visible light, but radio waves, ultraviolet light and X-rays are emitted, together with streams of charged particles (protons and electrons). These all have noticeable effects on the Earth. The X-rays and ultraviolet light cause disturbances in the ionosphere. This in turn disrupts radio communications. The ionosphere usually acts as a reflector for short-wave radio signals being transmitted on the Earth. A flare may cause a complete fade-out as the ionosphere temporarily absorbs the radio signals instead of reflecting them. It takes about two days for the streams of charged particles to reach the Earth. They may result in especially bright aurorae and can also cause fluctuations in the Earth's magnetic field.

The solar rotation

The rotation of the Sun is easily discerned by the passage of sunspots across the face of the Sun. The average rotation period is about 27 days, but the Sun does not rotate as a solid body. The rotation period varies with latitude, the period at the equator being shorter than that at higher latitudes. This is known as **differential rotation**. The Sun's axis of rotation is inclined to the plane of the Earth's orbit, so the apparent paths of sunspots across the face of the Sun vary during the year.

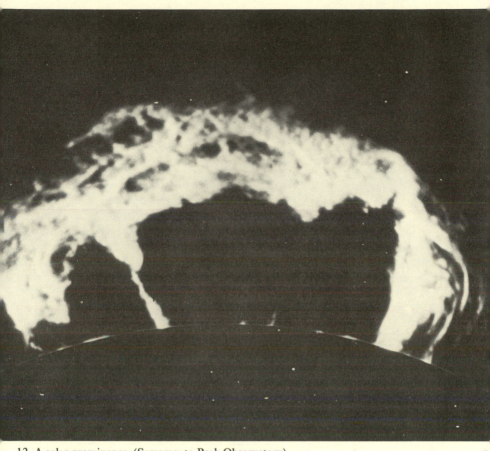

12. A solar prominence. (Sacramento Peak Observatory)

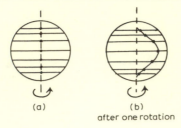

17.1 The differential rotation of the Sun. The equatorial regions rotate considerably faster than higher latitudes

(a)

(b)
after one rotation

The solar cycle

The same level of solar activity is not maintained the whole time, but it fluctuates on a cycle of about 11 years. At minimum there is virtually no activity. The amount of activity during a year is reflected in the **relative sunspot number** which takes into account the number and area of spots

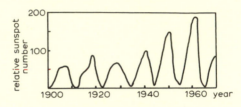

17.2 The sunspot cycle

which have appeared during the year. During the cycle, the solar latitude at which sunspots appear changes in a regular way. At the beginning of a cycle, new spots appear at high latitudes. As the cycle progresses, the new spots appear at lower and lower latitudes, until at the end they are found only near the equator. A graph on which all the sunspot latitudes are plotted against the date produces the characteristic **'Butterfly Diagram'**. Of course, the latitude of any individual spot does not change during its lifetime; it is the general location of spots which moves.

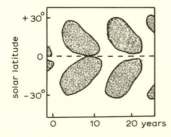

17.3 Schematic version of the solar 'Butterfly Diagram' showing the changing latitude of the appearance of sunspots through the solar cycle

The origin of solar activity

Solar activity is intimately connected with the Sun's magnetic field and differential rotation. Sunspots are areas of high magnetic field and often occur in pairs, one spot being a N-pole and the other a S-pole. It may be that energy is stored up as the differential rotation distorts the Sun's magnetic field, until the energy is released during the next burst of solar activity.

The chromosphere

The chromosphere is a thin outer layer of the Sun's atmosphere. It is seen as a reddish glow around the Sun during a total eclipse.

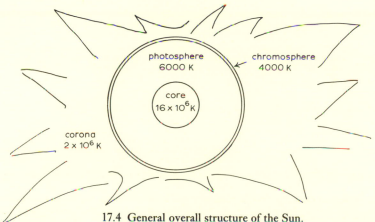

17.4 General overall structure of the Sun.
The temperature steadily decreases outwards from the core

The corona

The corona is a faint halo of gas surrounding the Sun. It extends to several times the diameter of the photosphere. Because it is so faint, it can only be seen during a total eclipse, or by the use of a special instrument called a coronagraph, which blots out the light from the photosphere. The gas particles in the corona have extremely high energies. If the speed of the particles is interpreted as a temperature, it puts the corona at over 1,000,000° K.

The solar wind

Rapidly moving particles are constantly streaming out from the Sun into interplanetary space. These are chiefly the charged particles electrons and protons. This stream is called 'the Solar Wind'.

TABLE 17.1 Properties of the Sun

radius	700,000 km (over 100 × Earth's radius)
mass	2×10^{30} kg (300,000 × Earth's mass)
average relative density	1·4 (compared with 1·0 for water and 5·5 for the Earth)
energy output	10^{23} kilowatts
surface temperature	5,800 K
distance	$1·5 \times 10^8$ km
age	5,000 million years
expected lifetime	10,000 million years

18 *Matter, light and energy*

Almost everything that is known about distant astronomical objects has been discovered from analysis and interpretation of the radiation which reaches us from them. The relationship between radiation and matter is therefore extremely important to astronomy.

The nature of light

Light is a form of energy. It is radiated in the form of tiny 'packets', called **photons**, which, in a vacuum, travel with the vast speed of 3×10^8 m/s. Each photon has associated with it a particular amount of energy, which determines the colour of the light. For example, in blue light the photons have more energy than in red light. However, if we assume that light behaves only as a stream of bullet-like particles it is not possible to explain adequately all the aspects of the behaviour of light. In many situations, the photons of light show *wave-like* behaviour. The two personalities of light are combined, and photons are described as 'wave-packets'. Sometimes the wave properties of light are more apparent; at other times the particle behaviour is more obvious. As light waves are oscillations of linked electric and magnetic fields, they are called **electromagnetic waves**.

18.1 Definition of wavelength, λ

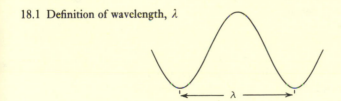

Important quantities for all waves are their **wavelength**, λ (the distance between two consecutive waves), their **frequency**, f (the number of waves sent out each second), and the velocity with which the waves travel (c is used to represent the velocity of light). These quantities are linked by the simple equation

$$\lambda f = c$$

The wavelength of light is so small that it is measured in a special unit. The Ångström (Å) is 10^{-8} cm, and red light has a wavelength around 7,000 Å. In the S.I. system of units, nanometers (nm) are used. 1 nm $= 10^{-9}$ m. 10 Å $= 1$ nm. The frequency of light waves is also related to the energy carried by the photons. The relation is $E = hf$, where h is a constant called Planck's constant.

The complete electromagnetic spectrum

If a beam of light, containing waves of different frequencies, is split up so that the different frequencies are separated out, the result is a **spectrum**. The process of breaking light down into its component frequencies is called **dispersion**. Dispersion is most readily achieved with a glass prism.

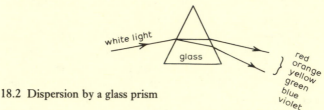

18.2 Dispersion by a glass prism

Our eyes can distinguish a whole range of frequencies which our brains interpret as the range of colours: red, orange, yellow, green, blue and violet.

Some confusion in understanding colours arises because of the interesting response our eyes have on receiving combinations of light of different frequencies. For example, if our eyes receive photons of red and green light simultaneously, then the sensation is that of seeing yellow, although no photons of yellow light may be arriving. So it is necessary to distinguish between impure colours which arise from combinations of other colours, and pure spectral colours which are of a very narrow band of frequency only.

The range of frequencies to which our eyes respond — what we call light — is in fact only a small part of the possible range of frequencies. The radiations outside the visible range of frequencies are all identical in nature to light and travel with the same speed; however, our eyes are not sensitive to them and they have some properties different from light. The entire range of radiations is termed the **complete electromagnetic spectrum**.

type of radiation	γ-rays	X-rays	UV	visible	IR	radio waves
typical wavelength	10^{-11} m	10^{-9} m	10^{-7} m	5×10^{-7} m	10^{-6} m	10^{-3} m $\sim$ 1m

18.3 The electromagnetic spectrum

Radiations with wavelengths shorter than light have higher frequencies and so each photon carries more energy. These radiations, of which X-rays are an example, tend to be dangerous to life because of the high energies of the photons which damage cells and tissue. To the long wavelength side of light lie infrared radiation and radio waves.

In astronomy, observation of radiations outside the visible range is increasingly important. The value of radio astronomy in extending our knowledge of the invisible Universe was established in the late 1940s; more recently, infrared, X-ray and gamma-ray astronomy have come into their own. A major difficulty has been that the Earth's upper atmosphere prevents all these radiations reaching ground level. Most of the ultraviolet and all of the X-rays and gamma-rays are absorbed. There is also a considerable amount of absorption in the infrared region, caused largely by water vapour. Progress has only been made by observations made from high-flying balloons and rockets or orbiting satellites. Some infrared observations can be made in desert areas or at high altitudes where the air is very dry.

The nature of matter

Matter is composed of the tiny particles we call **atoms**. There are about 92 different basic materials called **elements** each of which is composed of its own special kind of atom. Different types of atoms may combine chemically to form **molecules**. A material composed of molecules is a **compound**. With 92 naturally occurring, different elements, there is a large number of possible compounds.

The spacing of the atoms or molecules determines whether a material is solid, liquid, or gas, and in turn the spacing is influenced by the surrounding temperature and pressure. In a gas the molecules are virtually independent, widely spaced and moving about rapidly in random directions. In a solid the molecules are fixed in close positions, free only to vibrate. The liquid state is intermediate between these two extremes. Whatever state the material is in, an increase in temperature always results in the molecules moving or vibrating more rapidly. In fact, what we call temperature is really a measure of the energy of motion of the atoms and molecules.

Even an atom is not the smallest of particles. An atom consists of a **nucleus**, that carries a positive electric charge and a cloud of negatively charged **electrons** that are held to the nucleus by electrostatic attraction. The nucleus is extremely tiny (10^{-15} m) compared with the size of the atom as a whole (10^{-10} m).

The chief constituents of the nucleus are positively charged **protons**, and **neutrons** that carry no charge but contribute to the stability of the nucleus. Neutrons and protons are approximately the same size and each is nearly 2,000 times more massive than the electron. The electric charges carried by protons and electrons are equal in size, but opposite in sign. If a whole atom is to be electrically neutral, the number of electrons in the electron cloud equals the number of protons in the nucleus. Atoms of different elements are distinguished by the number of protons in the nucleus. Hydrogen, whose nucleus consists simply of one proton, is the lightest and simplest of the elements. More complex elements are built by the addition of protons, one by one, to the nucleus. If the nucleus is going to hold together, it also needs a certain number of neutrons to stabilize it. For any given number of protons, there may be several different possible numbers of neutrons which can create a stable nucleus. The different varieties of the same element which can be created by having different numbers of neutrons in the nucleus are called the **isotopes** of the element. The number of the protons in the nucleus (Z) is called the **atomic number** and the total number of neutrons and protons (A), the **atomic mass number**.

Line spectra

If the light from a glowing gas (for example, the red–glowing neon used in shop signs) is split up into its component colours by a spectrograph (an instrument for observing or photographing spectra) the result is a series of bright lines at particular colours, rather than the continuous blend of colours seen in the rainbow, for example.

The origin of line spectra like this can be explained if we understand how photons of light are emitted from atoms. The electrons of an atom possess energy because they are experiencing the electrostatic force of attraction from the positively charged nucleus. It turns out that nature only allows electrons in atoms to possess certain very particular values of energy. Furthermore, no two electrons in the same atom are allowed to have the same energy. The electrons are allowed to absorb energy and lose it again, as long as they always end up with a value which is allowed. When an electron loses a burst of energy, that energy is sent out in the form of a photon of light. In most usual situations, it is only one electron—the outermost one—which is free to absorb and release energy.

We describe the situation by saying that this outer electron has a number of possible energy levels. These can be imagined as unevenly spaced rungs on a ladder; the electron is allowed to jump from one rung to another. The correct term for such a jump is a **transition**. Each possible transition results in a photon of a particular precise energy, which means, of course, a particular colour. The spectrum of light from gas atoms therefore appears as a series of bright coloured lines with total blackness in between. Such a spectrum is an **emission line spectrum**. Each element has its own characteristic set of emission lines, so the spectrum is an important tool in the identification of the presence of particular elements, especially in distant astronomical objects.

One way in which an electron in an atom can become excited is to *absorb* light. Electrons will absorb photons if the energy of the photon corresponds to an energy jump that the electron is allowed to make. If light of all wavelengths, a **continuous spectrum**, is shone on to a cloud of gas, light is absorbed just at the same precise colours that the gas would emit. The appearance of the spectrum is then that of a continuous blend of all the spectral colours, crossed by sharp black lines. It is called an **absorption line spectrum**. The majority of stellar spectra are absorption line spectra.

The continuous spectrum

Line spectra are a consequence of the very precise energy levels which the electrons in a gas atom are allowed to have. In addition, all matter emits radiation over a continuous range of wavelengths. The actual amount of light varies with wavelength in a way which is governed by the temperature of the object.

Graphs that show the variation in light intensity with wavelength, as given off by objects at different temperatures, make a family of curves of similar shape, called Planck curves. The maximum point of a Planck curve occurs at shorter wavelengths, the hotter the object. A simple illustration of this fact is that as a poker heats up in a fire it glows red, then yellowish and finally blue–white as its temperature increases. Furthermore, the *total*

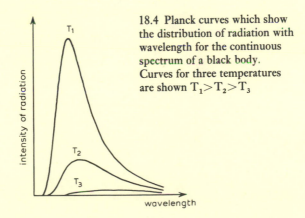

18.4 Planck curves which show the distribution of radiation with wavelength for the continuous spectrum of a black body. Curves for three temperatures are shown $T_1 > T_2 > T_3$

amount of light emitted increases sharply as the temperature goes up. The quantity of energy emitted as light is given by the equation $E = \sigma T^4$ where σ is a constant. This relation is known as the Stefan–Boltzmann Law.

The conditions in the interior of a star are such that the spectrum of light leaving the photosphere is continuous, and the distribution of light intensity with wavelength approximates very well to the Planck curve for the temperature of the star. However, the much less dense and cooler gases in the outer layers of the star (its atmosphere) superimpose the characteristic pattern of absorption lines in the continuous spectrum.

The doppler effect

If a source of light is moving towards or away from an observer, the light waves appear either compressed or stretched out respectively. This means that the observer sees light at a *different wavelength*, than if the source of light was stationary relative to him. If the light has a line spectrum, the effect is to shift the spectral lines to the red if the source is moving away, and to the blue if it is moving towards the observer. This is **the doppler effect** which is extremely useful in astronomy, for it means that it is possible to measure the velocities towards or away from us of stars and other bodies. Motion across the line of sight does not affect the spectral lines. The change in wavelength, $\Delta\lambda$, of a line of normal wavelength λ, caused by a relative velocity of v is given approximately by the very simple relation:

$$\frac{v}{c} = \frac{\Delta\lambda}{\lambda}$$

where c is the velocity of light.

19 *Stellar spectra*

The photosphere of a star is heated by radiation from inside and the light from it has a continuous spectrum. The distribution of light in the continuous spectrum has the approximate shape of a Planck curve. The surface temperature of the star may therefore be deduced by finding the Planck curve which fits the observed spectrum best. This is the temperature of the photosphere, and not the interior which is far hotter, of course.

Going outwards from the centre of a star, the gas is gradually cooler and less dense. The outermost layers (the star's atmosphere) impress on the continuous spectrum an absorption line spectrum. A few exceptional stars also show emission lines in their spectra due to unusual circumstances.

Classification of spectra

Stellar spectra are classified in what is essentially a temperature sequence. However, the spectral classes are still known by letters left over from an early classification scheme which was originally in alphabetical order, but which later turned out to be unhelpful.

class	typical temperature	colour	chief features in spectrum
O, B	20,000° K	blue-white	very few features—but including lines of helium
A	9,000° K	blue	Balmer lines of hydrogen and lines of ionized metals and neutral iron
F	7,000° K	blue-green	lines of hydrogen (weaker than in A), ionized and neutral metal lines
G	5,500° K	yellow	ionized and neutral metal lines and some molecular bands
K	4,000° K	orange	neutral metal lines and molecular bands
M	3,000° K	red	dominated by molecular bands

The order of spectral classes can easily be remembered from the mnemonic: O Be A Fine Girl, Kiss Me! Each class is also subdivided into 10 subclasses numbered 0 to 9. Thus, for example, the Sun is classified as

G2. Also, small numbers of the coolest stars are classified in three extra groups, R, N and S according to the detailed appearance of their spectra.

The differences in colour and spectral appearance of the majority of stars are almost entirely a result of the temperature differences. The colour depends on the distribution of radiation in the spectrum according to the Planck curve for the star's temperature.

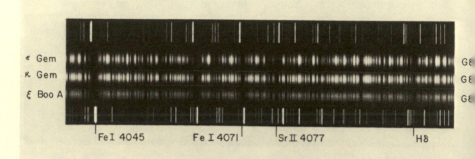

13. Typical absorption line spectra of stars. Above and below the three stellar spectra are emission line spectra of iron, for comparison. The three stars are all of spectral class G8, a little cooler than the Sun. ξ Boo A is a dwarf, κ Gem a giant and ε Gem a supergiant. Small differences can be detected between the three spectra which arise from their luminosity differences. (Lick Observatory)

The different appearances of the absorption line spectra of stars with different temperatures can easily be understood qualitatively. The conditions for given atoms (or molecules) to absorb and scatter light are quite critical. In the very hot stars all the hydrogen is ionized; there are no neutral hydrogen atoms to cause absorption and so the absorption lines of hydrogen are not seen. Many other elements also have several electrons stripped away and the resulting ions absorb very little in the visible part of the spectrum.

Moving down the temperature sequence we come to A and F stars, in which not all the atoms are ionized and which are hot enough to give the optimum conditions for hydrogen atoms to absorb light. In the cooler G and K stars the optimum conditions occur for absorption by metals and other heavier elements, chiefly iron, and it is not hot enough for many atoms to be ionized. In fact, even a few molecules can form without being torn apart by excess heat. In the coolest M stars the spectrum is dominated by the characteristic bands of molecular absorption lines.

The composition of stars

The appearance of characteristic lines of an element in the spectrum of a star shows that the element must be present in the star's atmosphere. Of course, the absence of lines does not imply the converse, because the temperature conditions might not be right to produce absorption lines. Astronomers have found that most stars have a composition very similar to the Sun's. About 75 per cent of a star's mass is hydrogen, around 25 per cent helium and less than 1 per cent of the mass is made up of most of the other elements. Comparatively few stable elements have never been found in stars. The spectrum of the star gives enough information to deduce the temperature and the quantities of different elements in the star's atmosphere, though in practice it can be quite difficult to separate out the effects of the two.

20 *The stars—physical properties*

Distances

The distance to a comparatively nearby star can be found by measuring its parallax; this is the apparent change in its angular position relative to more distant stars, as the Earth revolves in its orbit about the Sun. During the course of a year a nearby star seems to move in a tiny ellipse, around a second of arc or less across, against the background of distant stars. The distance of a star whose parallax is 1 arc sec is called 1 **parsec** (pc). 1 pc is approximately 3·25 light years. Of course, the smaller the parallax angle, the more distant the star.

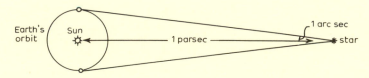

20.1 Stellar parallax and the definition of the parsec

In practice, observed parallaxes are extremely difficult to measure as they are so small, and they are only known accurately for a few hundred stars. The nearest star to the Sun, Proxima Centauri, is 4 light years away (i.e. about 1·3 pc) and there are only about 50 stars known to be within 5 pc of the Sun.

There are a number of indirect ways of measuring the distance to the more distant stars. The star's spectrum is classified and the spectral class tells the surface temperature of the star. For the majority of stars, there is a unique relationship between surface temperature and total energy output (see Section 21), so it is possible to deduce how much energy the star is actually emitting. When this is compared with the *apparent* brightness of the star, its distance can be judged. Another method can be employed for certain variable stars, called Cepheids, and this technique will be explained in the section on variable stars.

Proper motion and radial velocity

There is no obvious change in the relative positions of the stars either from night to night or even from year to year, and yet the stars are in fact moving. Over a timescale of a few tens of thousands of years, the constellations as we know them will be altered significantly, some of them beyond recognition. This true relative motion is called **proper motion**, and is measured in seconds of arc per year. The star with the largest known proper motion is a 13th magnitude object known as Barnard's Star, with a value of 10 seconds of arc per year. Proper motion is additional to the cyclical motion caused by annual parallax.

Proper motion is a measure of a star's angular velocity at right angles to our line of sight. Stars may also be moving towards or away from us. This component of their velocity relative to the Solar System is called their **radial velocity**. Radial velocities are much easier to find than proper motions as they show up through the doppler shifts in stellar spectra. The true space velocity can only be deduced if the star's distance is also known.

Absolute and apparent magnitudes

The **apparent magnitude** of a star is a measure of how bright the star seems to be. On the magnitude scale, the brightest stars are allotted the smallest numbers. The brightest stars in the sky are approximately of magnitude 1, whereas the faintest stars which can be seen without the aid of a telescope are about magnitude 6. Some of the planets are brighter than 1st magnitude stars so that their magnitudes are negative numbers.

The precise definition that relates magnitudes to the amount of light energy received from the stars is rather a strange one; it grew up primarily because of the response of the human eye. If two stars have magnitudes m_1 and m_2 and the light energy reaching us from the stars each second is respectively L_1 and L_2, the four quantities are related by

$$m_1 - m_2 = -2 \cdot 5 \left(\log_{10} \frac{L_1}{L_2} \right) \qquad (20.1)$$

The apparent magnitude, m, of a star depends both on its distance and on its intrinsic brightness. The **absolute magnitude**, M, is a measure of the intrinsic brightness and is defined as the magnitude the star would appear to have if it were at a distance of 10 pc from us. We can apply equation (20.1) to relate the apparent magnitude, m, and the absolute magnitude, M, of a star.

The amount of light received from any light source diminishes as the square of the distance from the source. If L_1 is used to represent the flux of light that would be received from a star if it were at a distance of 10 pc, and L_2 is the actual flux of light energy received, the inverse square law gives us

$$\frac{L_1}{L_2} = \frac{d^2}{100};$$

m_1 now represents M, and m_2 represents m. Making the substitutions in equation (20.1) and rearranging gives

$$m - M = 2 \cdot 5 \ (\log_{10} \frac{d^2}{100})$$
$$= 5 \log d - 5$$

Alternatively, as $d = 1/p$ where p is the parallax (in arc sec)

$$m - M = -(5 \log p + 5)$$

Radii

In a telescope all stars appear simply as points of light, so it is impossible to measure the size of a star directly. The radii of a few stars have been measured with an instrument called a stellar interferometer. Others can be deduced for stars that are members of binary systems. If the luminosity, distance and temperature of a star are known it is possible to deduce the size, as the amount of light given out per unit area of the star's surface can be calculated theoretically from these quantities, and so the surface area can be deduced.

The results show that there is a very large range in the sizes of the stars, from only one tenth to several hundred times the size of the Sun. The smaller stars are referred to as **dwarfs** while the large ones are called **giants**.

Masses

What is it that determines the temperature of a newly formed star? It seems clear that the temperature attained by a star depends almost entirely on the initial mass of the star. The larger the star, the hotter it is. It is not surprising that there is a straightforward relationship between the luminosity (or brightness) of a star, and its mass. The law seems to be $L \propto M^{3 \cdot 7}$. This result, however, is based on measurements of only a small number of stars, but the fact that the luminosity depends on a high power of the mass appears well established. Only for stars which are members of binary systems do we have any way of deducing the mass, and it is not even possible to do this for most binary systems! Consequently only a few dozen stars have had their mass determined accurately.

The masses of stars are most usually given in terms of the solar mass. The range of stellar masses seems to be between 0·1 and 50 solar masses. It is thought that if stars of more than about 50 solar masses try to form they explode before becoming proper stars. Objects below 0·1 solar mass are too small for energy producing nuclear reactions ever to start.

21 *The Hertzsprung–Russell diagram and stellar evolution*

Without doubt the **Hertzsprung-Russell diagram** is the most important graph ever plotted by astrophysicists. To form a Hertzsprung-Russell diagram the total energy output of each of a group of stars is plotted against the corresponding stellar temperature. In practice, the absolute magnitude is plotted rather than energy output since the former is more readily measurable by existing astronomical instruments. However, the magnitude and energy are directly related to each other. Occasionally the sequence of spectral classes is used in preference to temperature; again, spectral class is easier to determine than temperature.

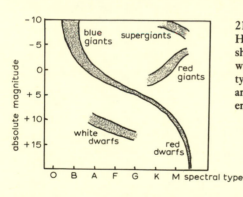

21.1 Schematic version of the Hertzsprung-Russell diagram showing the principal regions in which stars are found. Spectral type is a measure of temperature, and absolute magnitude of the energy output

One might expect that stars would be randomly distributed in such a plot but this is far from the case. Most stars are located along a narrow strip known as the 'Main Sequence'. Because of this fact, the Hertzsprung-Russell diagram has become an important tool in understanding stellar evolution, and star clusters.

The main sequence sweeps diagonally from the upper left to the lower right. On the basis of what we now know, it is clear that the main sequence represents, from top to bottom, decreasing mass, luminosity and temperature and a colour sequence from blue to red. The massive stars at the top end of the main sequence are **blue giants**; those lower down the sequence are the dwarfs.

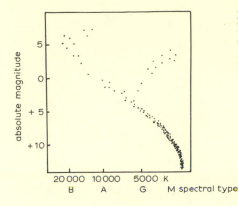

21.2 Schematic Hertzsprung-Russell diagram of stars in the solar neighbourhood. The majority are faint red dwarfs on the main sequence

Not *all* stars, however, lie on the main sequence. To the right of the main sequence lie stars which are much more luminous than main sequence stars of the same temperature. This is because they are much larger, and so they are called **red giants**. Even larger stars are known; these are the **supergiants**. Below the main sequence is a region of faint stars called **white dwarfs**. Some white dwarfs belong to binary systems, so it has been discovered that their masses are in the range one to one-half of a solar mass. However, these stars are very much smaller than the Sun, only one-hundredth the diameter, which means that their density is incredibly high. In fact, the density inside a white dwarf is so great that even the atoms are crushed. The electrons, which normally form a shell around the atomic nucleus, are pushed closer to the nuclei to form a continuous sea of electrons. Matter in this state is said to be **degenerate**.

Evolution of stars

It is believed that white dwarfs and red giants were once ordinary stars which were on the main sequence, but as time went on the continuing production of energy caused the stars to change. Stars have a finite lifetime. They are born, evolve and eventually die.

It is now accepted that stars form from the gas and dust which exists in space between the stars. Once the contraction of a gas cloud has started, the mutual gravitational pull of the particles rapidly accelerates the collapse of the **protostar**, heating up the interior until the temperature at the centre reaches 1 million degrees. When this point is reached, nuclear reactions, the source of the star's energy, can begin.

The composition of the material from which stars have formed is predominantly hydrogen; the Sun's mass, for example, is three-quarters hydrogen. Hydrogen is the chief fuel which stars have to 'burn' in the nuclear processes which keep them shining.

In many stars, including the Sun, one of the simplest and most important nuclear processes is the conversion of hydrogen to helium in a series of reactions called the **proton-proton chain**. This serves as a good example of the many possible reactions which can occur in stars. The nucleus of a hydrogen atom is just a single proton. When two protons collide they may fuse together and so form a **deuteron**, that is the nucleus of what is sometimes called 'heavy hydrogen', which consists of a proton and neutron. In addition, a positively charged positron and a neutrino are released, along with some energy. A positron is similar to an electron, but carrying a positive electric charge. A neutrino is an elusive particle which has neither mass nor electric charge, but carries energy at the speed of light. This reaction can be symbolized by an equation:

$$_1^1H + {}_1^1H \longrightarrow {}_1^2H + e^+ + v + \text{energy},$$

or,   proton + proton $\longrightarrow$ deuteron + positron + neutrino

The deuterons which are produced rapidly undergo another reaction in which they collide with another proton to form an isotope of helium which is short of a neutron

$$_1^1H + {}_1^2H \longrightarrow {}_2^3He + \text{energy},$$

or,   proton + deuteron $\longrightarrow$ 'light' helium

The light helium nuclei then react with one another to form normal helium, which has two protons and two neutrons:

$$_2^3He + {}_2^3He \longrightarrow {}_2^4He + {}_1^1H + {}_1^1H + \text{energy}$$

In the process two protons are set free again.

All three parts of this chain reaction are accompanied by the release of energy, energy which finally leaves the surface of the star as heat and light. The energy actually comes from the annihilation of matter. Mass is being converted into energy. The Sun, for example, loses 4 million tonnes each second!

A star that is consuming its hydrogen fuel in a normal manner will be found on the main sequence in the Hertzsprung-Russell diagram. However, eventually a significant proportion of the star's core will be converted to helium, and then changes occur in the star's appearance. The outer parts of the star expand to form a tenuous envelope, but at the same time the core contracts until the central temperature goes up even further. These changes cause the star to become redder, yet more luminous. The star moves to the right of the main sequence and becomes a red giant.

The *rate* at which a star uses up its hydrogen depends on its initial mass. The most massive stars evolve fastest, while dwarf stars such as the Sun survive longest. The main sequence lifetime for a hot blue O-type star with a mass 35 times the Sun's is only about four million years (4×10^6 yr), whereas the Sun will stay near the main sequence for ten thousand million years (10^{10} yr). Striking proof of this strong dependence of lifetime on mass is contained in Hertzsprung-Russell diagrams plotted for star clusters. It is fair to assume that all the stars that belong to a cluster were formed at about the same time, at least by astronomical standards. Because the rate of evolution increases with mass, the stars near the top of the main sequence evolve the most rapidly and so moved to the right furthest towards the giant branch, thus giving the main sequence a distinct curve. In fact, the older the cluster, the further down the main sequence the curving (or turnoff) starts. This enables astronomers to deduce the relative ages of star clusters.

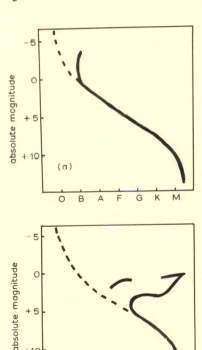

21.3 Schematic Hertzsprung-Russell diagrams for (a) a young star cluster and (b) an old star cluster. In each case, the main sequence for completely unevolved stars is shown as a dotted line

As a star exhausts the supplies of hydrogen in its interior a readjustment of structure becomes necessary. The central regions of the star shrink, and in doing so release energy. This energy heats the outer layers of the star. In these layers hydrogen may burn in a shell around the star, releasing more nuclear energy. The net result is that the stellar atmosphere becomes enormously distended. The star is now a red giant, having evolved away from the main sequence. After the red giant phase further contraction of the core takes place generating higher temperatures in the interior which makes it possible for new nuclear reactions to occur. In fact, a whole sequence of contractions occur in which the interior gets hotter and hotter, and different nuclear reactions convert the stellar material into heavier and heavier elements. Eventually, the star becomes bluer again and returns to the vicinity of the main sequence. In doing so, it passes through a stage of instability when it may be a **pulsating variable star**. Now it is a dying star on its way to the white dwarf graveyard where it will gradually cool down and fade out. It will then cease to be a luminous star.

It was noted earlier that white dwarfs do not have masses much greater than the Sun's, so what happens to the more massive stars? It is thought that these end their life suddenly in a gigantic explosion, called a **supernova**. Supernovae are not common events. The last one observed in our own Galaxy was seen by Kepler in 1604. They probably occur at a rate of 1 per 50 years in the Milky Way, but most of them are not seen because they are hidden by dense clouds of dust. From time to time they are observed in other galaxies. So enormous is the explosion that the supernova often becomes comparable in brightness to all the other stars in the galaxy put together! The supernova throws off a cloud of gas called a supernova remnant. The Crab Nebula in Taurus is the remains of a supernova which occurred in 1054. We know this because of the careful records kept by the Chinese astronomers.

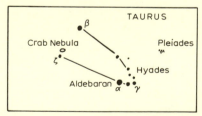

21.4 Location of the Crab Nebula (about 11th mag.)

The central star which is left behind after the explosion is so collapsed that it may become even denser than a white dwarf. It is called a **neutron star**, because the electrons and protons which would normally form atoms

14. The Crab Nebula. (Hale Observatories)

are squashed together to form neutrons under the enormous pressure. The diameter of a neutron star is about 10 km, and its density is that of the nucleus of an atom, around 10^{15} times higher than the density of ordinary matter. Neutron stars have been identified now with **pulsars**. Pulsars were discovered in 1967 because of the strange pattern of radio waves that they emit. Sharp pulses of radio waves occur at very regular intervals. For different pulsars the interval ranges between a few seconds and a small fraction of a second. The pulses are so regular that at first the possibility that these signals were coming from intelligent creatures was not discounted! The central star of the Crab Nebula is a pulsar, the only one that has been seen optically, and, furthermore, the light and X-rays from this star flash in time with the radio pulses. Theorists have found that it is quite possible to explain the pulses from pulsars in terms of the rapid rotation of an extremely dense object such as a neutron star. For the Crab Nebula pulsar the rotation rate is 30 times per second.

It is thought that the mass of a neutron star cannot exceed 2 solar masses; theoretical astronomers have shown that a neutron star any heavier than this will be crushed into an ever-decreasing volume by its own gravity. As the star shrinks the gravitational field strength $(g = GM/r^2)$ continues to increase and so the escape velocity at its surface $(v = \sqrt{2GM/r})$ gets larger. Eventually the escape velocity equals the speed of light. This has the intriguing consequence that nothing whatever can escape from the collapsed star by any means. Such a star is therefore totally invisible and is known as a **black hole**. It has been tentatively suggested that a few binary star systems contain a black hole; these particular systems are very powerful X-ray sources.

22 Double stars

Perhaps 50 per cent of all stars, certainly a significant proportion, are members of binary systems. True binary stars must not be confused with **optical doubles,** in which a chance alignment causes two stars at different distances to appear very close in the sky. Alcor and Mizar in Ursa Major form a well-known optical double. True binaries are physically associated and are in orbit about each other. The rules which govern their orbits are exactly the same as those which Kepler deduced for the motion of the planets around the Sun. In the case of a binary star system, each star is orbiting around the centre of gravity of the pair. The centre of gravity is simply the point of balance between the stars.

Visual binaries

Visual binaries are pairs of stars whose relative orbital motion can be seen over a period of time by telescopic observation. If the stars are far enough apart to be resolved by a telescope, the period is usually quite long, of the order of tens of years at least.

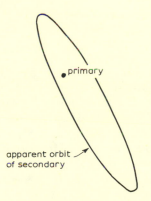

22.1 The apparent relative motion of the visual binary system α Centauri. The primary does not lie at one of the foci of the secondary's apparent orbit because of the angle at which the system is being viewed

Spectroscopic binaries

Some stars that appear to be single, even in the largest telescopes, exhibit changes in their spectra that show that they are really double stars. The spectra of the two stars are superimposed, but as one star moves towards us and the other away, the doppler effect causes the apparent wavelengths of the spectral lines of the two stars to shift in opposite directions, so the lines appear double. When both stars are moving across the line of sight of the observer, the lines are once again exactly superimposed. So, during the orbital period of the stars, there is a cycle of change in the spectrum. The shift of the spectrum lines can be measured and turned into the radial

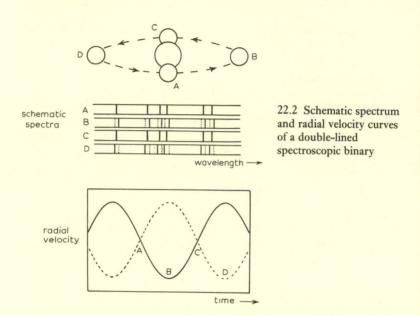

schematic spectra

22.2 Schematic spectrum and radial velocity curves of a double-lined spectroscopic binary

velocity (i.e. the velocity along the line of sight). If a series of these is plotted against time in a graph, the result is a radial velocity curve. From such a curve a limited amount of data can be deduced about the masses and separation of the stars. If the spectra of both stars can be seen as just described, the spectroscopic binary is said to be double-lined. Sometimes, one of the stars is much fainter than the other, so its presence is only detected by the periodic shift in the spectral lines of the other. This type of spectroscopic binary is said to be single-lined.

Eclipsing binaries

The orbits of binary stars may be tilted at any angle in space, and we have no way of finding out that angle, so it is not possible to find the masses of the stars precisely. However, it is easy to recognize binaries in which we are seeing the orbital planes exactly edge-on, because such stars go through a cycle of eclipses, as one star passes alternately in front of and behind the other. This means that the total apparent brightness of the binary changes regularly. The most famous example of such an **eclipsing binary** is Algol (β Persei). For eclipsing binaries, information can be deduced about the size, shape and masses of the stars.

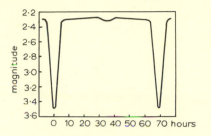

22.3 The light curve of the eclipsing binary Algol (β Persei)

Kepler's law for binary stars

The motion of the stars in a binary pair obeys the Law of Gravity, just as the motion of the planets around the Sun does. It follows that Kepler's Laws of planetary motion can be applied to binary stars too, with a little modification. The difference arises because the two stars in a binary pair usually have quite comparable masses, so the balance point between them lies somewhere in space along the imaginary line joining them. The masses of the planets are all very tiny compared with the mass of the Sun, however, so the balancing point between the Sun and one of its planets lies close to the centre of the Sun. That is why we can say that the planets travel around the Sun.

In the case of a binary star system, Kepler's First Law becomes, "The orbit of each star is an ellipse, with the centre of gravity of the system at one of the foci." The Third Law becomes, "The square of the period is proportional to the cube of the mean distance apart." The two stars swing about the centre of gravity in such a way that there is always one star on either side of it, so both stars take the same time to complete their orbits. Kepler's Third Law explains why visual binaries, which have to be wide apart to be resolvable in a telescope, always have long orbital periods.

To find V in terms of P, use distance = speed × time

$$2\pi a = V \times P,$$

$$F = G\frac{mM}{r^2} = \frac{MV^2}{a},$$

$$\frac{Gm}{r^2} = \frac{4\pi^2 a}{P^2 a},$$

$$\frac{Gm}{r^2} = \frac{4\pi^2 mr}{P^2(M+m)},$$

or

$$P^2 = \frac{4\pi^2}{G(M+m)}r^3,$$

and hence

$$P^2 \propto r^3.$$

in the case of a planet of mass m orbiting around the Sun (mass M), m is so much smaller than M that it can be ignored for simple calculations, and then you will see we have exactly the same equation as before (Section 16).

23 *Intrinsic variable stars*

Eclipsing binaries are one example of variable stars. However, true variable stars change in brightness because of internal changes in the star, and so are **intrinsic variables**. These variables are mostly stars in later stages of their evolution. Many different classes of variable stars have been identified. **Periodic variables** change in brightness with a very regular rhythm. **Semi-regular variables** are roughly periodic, though showing a certain amount of unpredictable variation, and **irregular variables** show no particular pattern in their variations.

Periodic variables

Most periodic variables are giants or supergiants. Their light output varies because the stars are actually pulsing in and out, with a consequent change in brightness as the surface area shrinks and grows. As the stars pulsate, the motion of the outer layers can be detected from the doppler shift of lines in their spectra. One of the first kind of pulsating variables to be studied was the group called **Cepheid variables**, which are named after Delta Cephei. The periods of variation of Cepheids are between about 3 and 100 days.

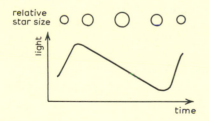

23.1 The pulsation of a Cepheid variable and the corresponding light curve

One of the most important discoveries concerning Cepheids was that the periods and the absolute magnitudes of these stars are directly related. This was first pointed out in 1910 by Miss H. Leavitt at Harvard, who was studying Cepheids in the Magellanic Clouds. (The Magellanic Clouds are small, nearby galaxies which can be seen only from southern latitudes on the Earth.) The importance of this period–magnitude relation is that Cepheids can be used for measuring distances. It is necessary to measure the period of variation and the apparent magnitude of a Cepheid; the absolute magnitude can be calculated from the relationship with period. Then, the distance can easily be found when both the apparent and absolute magnitudes are known. The chief difficulty at first was establishing exactly the absolute magnitude scale for Cepheids, because the distances to at least one or two Cepheids had to be found by an independent method. None are near enough to have a measurable parallax. However, by various stages, it has been possible to find the distances of some star clusters that contain Cepheids, and thus calibrate the period–luminosity law. The intrinsic brightnesses of Cepheids are so large that they can be identified in a number of nearby galaxies, whose distances can therefore be determined.

Another well-known class of periodic variables is the RR Lyrae stars. These are similar to Cepheids, but they are rather fainter and they have periods of less than one day. All RR Lyrae stars are of about the same intrinsic brightness but the light curves of stars of different periods have different shapes. They, like the Cepheids, are valuable 'standard candles' which can be used to survey the distances to the galaxies.

Long-period variables

These stars are all cool red giants or supergiants. The period of variation is often a year or more and the change in visual magnitude may be very large, amounting to several magnitudes. One famous example is Omicron Ceti, called Mira (wonderful), which varies in brightness between 9th and 4th magnitudes in a period of about a year. At minimum light, these stars have become substantially cooler than they are at maximum. As they are so red, the light output moves to the infrared region of the spectrum which cannot, of course, be seen by eye, so the visual magnitude drops sharply. What is happening in these stars is not fully understood, but their complex spectra suggests that they also may be pulsating.

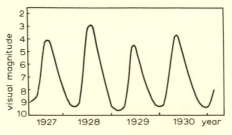

23.2 The light curve of Mira Ceti

Semi-regular and irregular variables

Semi-regular variables are somewhat similar to the long-period variables as they also are cool red giants or supergiants. However, the changes are usually rather smaller. The red supergiant Betelgeuse in Orion is a typical example. It varies by around 1 magnitude in 400 days or so.

There is a group of stars which suffer completely unpredictable changes in brightness. This might be either a sudden temporary increase or decrease in brightness. A famous example of this type is R Coronae Borealis which occasionally fades from 6th magnitude to 12th or 14th!

A further important group of irregular variables is that of T Tauri stars. These are newly formed stars that are still readjusting themselves before settling down on the main sequence. They are surrounded by dust shells that may eventually turn into systems of planets.

Explosive variables

We have already seen that some massive stars end their lives in a gigantic explosion called a supernova. There are also stars that suffer explosive outbursts of rather a small scale, but apparently without drastic consequences. These stars are called novae. During the explosion, the brightness may increase by about 13 magnitudes and a shell of gas is blown off. However, over several years the nova usually fades back to its original condition. Some novae have recurrent outbursts, either regularly or irregularly. All novae are members of close binary systems, in which the presence of a white dwarf triggers the nova outburst.

24 *The Milky Way*

On a clear, dark night it is easy to pick out the broad band of faint light which sweeps round the whole sky, through the constellations Sagittarius, Scutum, Cygnus, Cassiopeia, Perseus, Auriga, Crux Australis, Centaurus, Lupus and Scorpius. This is our view, from the inside, of the system of myriads of stars to which the Sun belongs — our Galaxy or the Milky Way.

The shape of the Milky Way

The Galaxy is shaped roughly like a disc with a central bulge. The Sun is situated in this disc, rather towards the edge. The centre of the Galaxy lies in the direction of Sagittarius, where extremely dense clouds of dust, gas and stars can be seen. Within the Galaxy, stars are not uniformly distributed. Towards the centre, the stars are much more crowded, and in the outer parts of the disc the Galaxy has spiral arms, in which stars and gas tend to be concentrated. The spiral arms have been mapped out largely through radio astronomy, but they are a feature found in many other galaxies. Also, stars are often found associated in groups called **clusters**.

Studies of the motions of stars show that the Galaxy is rotating about its centre. In fact, if this were not the case the pull of gravity would cause all the stars to fall to the centre! However, the Galaxy does not rotate as a rigid disc, but each star is in its own orbit round the centre. It will take the Sun 224 million years to make one revolution round the Galaxy. The central regions of the Galaxy rotate more rapidly than the outer regions, just as the inner planets of the Solar System circuit the Sun fastest.

Stars are not the only constituent of the Milky Way. Between the stars lie clouds of gas and dust which are an important part of the structure of the Galaxy.

Interstellar matter

Careful observation of the Milky Way with the naked eye shows that there are dark patches and lanes crossing the clouds of stars. These are not regions where there are fewer stars, but clouds of dark, obscuring dust. A good example is the Coal Sack near to the Southern Cross. As well as these dense clouds of dust, absorbing matter seems to pervade most of interstellar space. One of the chief effects this has is to make the light from

15. The Pleiades, showing the reflexion nebulosity. (Lick Observatory)

16. The Horsehead, a famous dark nebula in the constellation of Orion. It shows up clearly against the background of a bright emission nebula. (Tautenberg Observatory, DDR)

distant stars look redder than it really is, like the effect fog sometimes has on the appearance of the headlights of an oncoming car. Certain of the absorption lines in the spectra of some distant stars have been identified as originating in the interstellar matter.

Not all interstellar clouds, or nebulae, are dark. Quite often, clouds of gas will either shine by reflected light or emit light of their own. **Reflexion nebulae** are clouds of dust and gas which shine with light reflected from nearby bright stars. A good example is the nebulosity surrounding the Pleiades. We can tell that the light is reflected because the spectrum of the light has the same absorption lines as the nearby stars.

Emission nebulae are gas clouds which absorb ultraviolet radiation from very hot nearby stars, and then re-emit light, so that the spectrum shows emission lines, chiefly those of hydrogen. One of the most famous examples is the Great Nebula in Orion (M42) easily seen with binoculars or a small telescope. It is situated in Orion's sword. Emission nebulae are usually associated with stars of spectral types O and B.

17. The Great Nebula in Orion. (Lick Observatory)

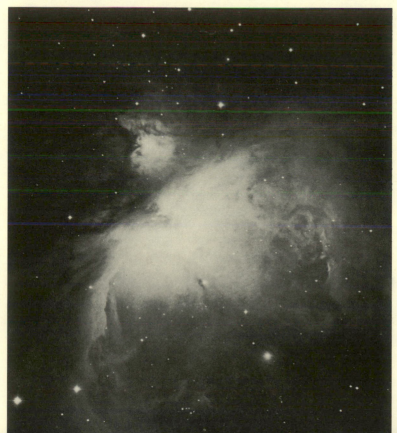

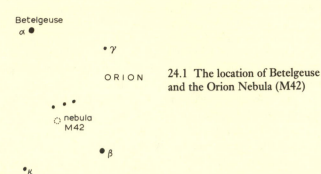

24.1 The location of Betelgeuse
and the Orion Nebula (M42)

Hydrogen gas emits light when it is excited by hot stars nearby, but until the advent of radio astronomy there was no way of finding out whether there were unexcited atoms of neutral hydrogen elsewhere. However, vast clouds of neutral hydrogen in space send out radio waves at a wavelength of 21 cm. In the early 1950s the 21-cm line was detected, particularly from the regions of the Milky Way. At first it was seen as an emission line, but later it was also detected as an absorption line in the radio spectra of more distant strong radio sources.

The discovery of the 21-cm line of neutral hydrogen heralded a great breakthrough in mapping the spiral arms of our Galaxy. The hydrogen is concentrated in the spiral arms. Because these arms are moving at different rates, the 21-cm radio waves are doppler-shifted by different amounts for the various arms. Clever detection work on the 21-cm spectrum has resulted in the mapping of the spiral arms out to great distances from the Sun. The radio waves can penetrate regions of space where the starlight is all absorbed, and optical astronomy is of no use.

The latest chapter in the discovery of interstellar matter is the detection of over 30 polyatomic molecules such as ammonia (NH_3) and formaldehyde (HCHO) and even more complex compounds. These have been found because they have transitions in the microwave region of the spectrum. They are mainly located in gas clouds such as the Orion Nebula and in the central part of the Galaxy.

Star clusters

There are two main types of star clusters: the **open (or galactic) clusters** and **globular clusters**. In addition, there are looser groupings of stars called **associations**.

A number of open clusters can easily be seen with binoculars or a small telescope, for example the Pleiades and the Hyades, both in Taurus, Praesepe in Cancer and the small double cluster h and χ Persei. Several hundred are known to exist and there are probably several thousand in the Milky Way, each containing a few hundred stars. The stars are well spaced and generally all resolvable. A galactic cluster has no particular shape.

18. The twin open clusters, h and χ Persei. (Royal Observatory Edinburgh)

On the other hand, globular clusters are densely packed spheres of stars containing vastly more stars than galactic clusters. About a hundred of these are known in our Galaxy. The brightest ones, Omega Centauri and 47 Tucanae, are in the southern hemisphere and can be seen with the naked eye. The best ones in northern skies are M13 in Hercules and M5 in Serpens, both about 6th magnitude and easily visible with a small telescope.

Galactic and globular clusters differ both in the kind of stars they contain and in their distribution in space. Galactic clusters are comparatively young, whereas globular clusters contain only old stars. The age of a cluster can be deduced by plotting the Hertzsprung-Russell diagram for it (see Section 21). The galactic clusters are all found close to

19. The globular cluster ω Centauri. (Anglo-Australian Observatory)

the plane of the Galaxy (hence their name) but the globular clusters are distributed in a sphere (or halo) surrounding the Galaxy, so many of them are a great distance from the disc. This distribution is probably a consequence of the immense age of these clusters, which formed before our Galaxy became a flat disc, some 12–15 thousand million years ago.

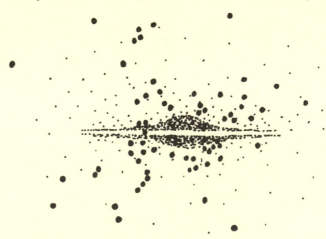

24.2 The distribution of globular clusters in our Galaxy, shown schematically

25 *The Universe of galaxies*

Before their true nature was understood, all misty patches in the sky were termed nebulae, and controversies raged about whether they were composed of stars or gas, and whether or not they were in the Milky Way. Of course, it turned out that some are gas clouds and some are star clusters in our own Galaxy, but others are distant galaxies in their own right. Larger and larger telescopes and better photography have revealed that the Universe is populated with vast numbers of galaxies, for as far as we can probe the depths of space.

Classification of galaxies

Like our own Milky Way, many galaxies (17 per cent) have spiral arms and are therefore known as spiral galaxies. Other galaxies (80 per cent) are elliptical in shape with no obvious structure. A few per cent of all galaxies have no regular shape at all. Of course, we do not know the angle at which we are viewing a galaxy, and consequently we cannot be sure of the exact shape. However, galaxies can be classified according to their general appearance by means of a simple system introduced by E. Hubble in 1925.

25.1 Hubble's scheme for the classification of galaxies

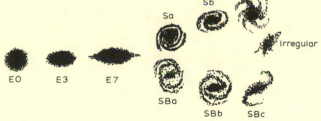

Elliptical galaxies (E) are smooth and structureless but vary in appearance from round (E0) to extended ellipses (E7). Spiral galaxies (S) have a bright nucleus from which emerge the spiral arms. In some spirals the arms start at the ends of a bright bar across the nucleus, and these are called barred spirals (SB). Spiral and barred spiral galaxies can be further

20. The spiral galaxy M83. (Anglo-Australian Observatory)

21. A distinctively barred spiral galaxy, NGC 1365. (Anglo-Australian
Observatory)

22. The irregular galaxy, M82. (Tautenberg Observatory, DDR)

subdivided into classes a, b, c and d, according to the increasing degree of development of the spiral arms. Intermediate between the elliptical and spiral galaxies are type S0 which show a definite nucleus, but little evidence of spiral arms. Galaxies of no particular form are classed as irregular (Irr).

Not all galaxies slot easily into this classification, and more complicated schemes are also used.

Distribution of galaxies in space

Photographs taken with large telescopes show that galaxies can be seen down to the limits of detection. They do not usually appear as isolated individuals, but in clusters. For example, there are comparatively nearby clusters of galaxies in the constellations Virgo, Coma Berenices and Hercules. Our Galaxy belongs to a cluster of galaxies that includes the Magellanic Clouds, which are irregular dwarf galaxies, and the great spiral galaxies in Andromeda (M31) and Triangulum (M33).

It is, of course, quite difficult to measure the distances to the galaxies, and astronomers use many tricks to arrive at estimates. For the nearest galaxies it is possible to identify Cepheid variables and other stars whose absolute magnitudes are known from prototypes in our own Galaxy. Then the distance can be deduced from the apparent magnitudes of these stars. For more distant galaxies, estimates have to be made on the basis of the apparent brightness and diameter of the galaxy as a whole.

The expansion of the Universe

The velocities of galaxies along our line of sight (radial velocities) can be found in exactly the same way as for stars: from the doppler shift of the spectral lines. E. Hubble, in 1929, showed that the galaxies are receding, and that there is a relationship between the speeds of recession of the galaxies and their distances. This relationship has come to be known as Hubble's Law, and can be written simply:

$$V = H \times D$$

where V is the velocity of recession of a galaxy at distance D, and H is a constant of proportionality. Recent estimates put the value of H at about 50–100 km/s per megaparsec of distance from us.

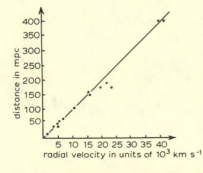

25.2 Hubble's Law. Each point represents a cluster of galaxies. The line corresponds to a Hubble constant of $100 \text{ km s}^{-1} \text{ mpc}^{-1}$

23. A cluster of galaxies called Abell 1060. The two bright images with spikes are nearby stars. The three brightest galaxies are ellipticals, showing no sign of spiral structure, but many spirals can also be picked out. (Anglo–Australian Observatory)

The fact that all galaxies are receding from us does not mean that we are in a privileged position in the Universe! To see this, imagine many dots on the surface of a balloon which is only partially blown up. As the balloon is inflated further, whichever dot you fix your attention on, all the others seem to be getting further from it, even though no dot is at the centre of expansion. The Universe is experiencing a similar general expansion.

Hubble's Law holds well for the galaxies whose distances can be measured independently, but these are only nearby galaxies. Can we assume that the law holds for more distant galaxies? If so, it is only necessary to measure the doppler shifts in the spectra of distant galaxies, then Hubble's Law can be used to find their distances. For objects moving away, the doppler shift is always to the red end of the spectrum, so the term redshift is used. The spectra of most distant galaxies that have been studied are redshifted by a factor of 0·7 in wavelength. These objects are thought to be about 2,000 megaparsecs from us; their light has journeyed through the vast reaches of intergalactic space for 6,500 million years, since before our Sun was born, in order to reach our telescopes.

In the early 1950s radio astronomers noticed that some elliptical galaxies are extremely powerful transmitters of radio waves. These are called radio galaxies, and their energy output in radio waves exceeds the luminosity in visible light. A great catastrophe in the centre of a galaxy probably gives rise to the radio emission.

There was no reason to suppose that the redshift law did not hold until, perhaps, the discovery of quasars. Quasi-stellar objects (quasar or QSO for short) were discovered in 1962, originally by radio astronomers. The radio waves that a quasar sends out are very similar to those emitted by a radio galaxy, but photographs show only a point of light, like a star, rather than a galaxy. Furthermore, many of these objects turned out to have redshifts so large (nearly 4) that Hubble's Law would make them the most distant objects known! The question then arises, if they are so far away, and yet we can see them clearly, how are they putting out so much light energy? Astronomers are still trying to find the answers to the many problems posed by quasars.

26 Telescopes

It is impossible to imagine what astronomers would have done without telescopes. The invention of the telescope in around 1600 meant a complete revolution in observational astronomy. It is still the most important tool, because it enables an astronomer to collect far more light from distant stars and galaxies than he can with the naked eye. This makes it possible to detect detail on nearer bodies, such as the planets, and to photograph extremely faint objects in the furthest depths of space.

Optics of simple telescopes

The optical system of any telescope has two basic parts: the **objective** and the **eyepiece**. The purpose of the objective is to collect light, so the larger its area, the fainter the objects that can be seen. In a **refracting telescope** the objective is a convex lens, whereas a **reflecting telescope** uses a curved mirror. The objective brings light to a focus. The eyepiece is like a magnifying glass which is used to view the focused image, and in fact, it is often composed of two or more lenses. The eyepiece might be replaced by a camera or by a spectrograph if the telescope is to be used to take photographs or spectrograms.

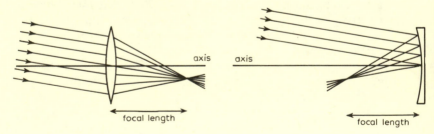

26.1 Beams of parallel light brought to foci by
(a) a convex lens
(b) a concave mirror

In a refracting telescope the objective lens and eyepiece are supported at either end of a long, closed tube. There are a number of different ways of arranging the viewing with a reflecting telescope. Small reflectors usually have a small flat mirror to send the image out to the side. This is called the

Newtonian system. In extremely large telescopes the observer may be able to sit in a small cage at the **prime focus. Cassegrain** reflectors have a small hole in the main mirror through which the light is reflected by a curved secondary mirror. This arrangement has the advantage that equipment can be mounted at the end of the telescope. If the equipment is very large, as many spectrographs are, even that might be impossible. This difficulty is overcome in the **coudé** system: the focus of the telescope is at a fixed point — often in a room below the main observatory, so that the image falls in exactly the same place wherever the telescope points in the sky. Surprisingly enough, secondary mirrors or even a prime focus cage do not significantly affect the performance of the telescope so long as they are not too large. Reflectors often have open tubes, which are just frameworks to hold the mirrors in place. This allows the free circulation of air round the mirrors which helps to improve the quality of the image. A solid tube, closed at one end only, tends to create a regular current of air which make the images dance about.

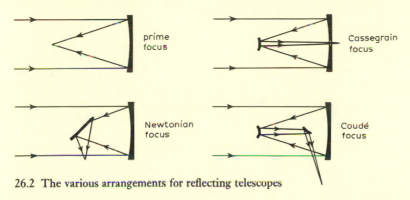

26.2 The various arrangements for reflecting telescopes

Angular magnification

The magnifying power of a telescope can be altered by use of different eyepieces. With an astronomical telescope one is viewing objects whose real size is unknown. What we do know, however, is the apparent *angular* size, as the angle subtended at the eye by the object is termed. The telescope makes an object appear nearer, and so increases the apparent angular size. Suppose the apparent angular size of a planet as seen with the naked eye is α, but with a telescope is β. The **angular magnification** of the telescope, m, is given by

$$m = \frac{\beta}{\alpha}$$

This quantity is in turn related to the focal lengths of the objective and eyepiece (f_o and f_e)

so

$$m = \frac{\beta}{\alpha} = \frac{f_o}{f_e}$$

For any particular telescope f_o is fixed, and is usually quite large, but various eyepieces of different f_e can be used to achieve a range of magnifications.

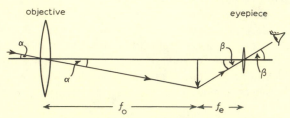

26.3 The angular magnification of a telescope

Resolution

One reason for having the telescope objective as large as possible is to have maximum light gathering power. However, the size of the objective also affects the **resolution** which is the ability of a telescope to distinguish two nearby objects or features. Even a point source such as a star forms an image which has a disc of definite size. This is not the true disc of the star,

26.4 The meaning of resolution

but an optical effect due to the bending (or diffraction) of light waves as they pass through the objective aperture. The disc may be surrounded by faint rings. The larger the aperture, the smaller the image disc, and hence the ability of the telescope to resolve close objects is improved.

Reflectors versus refractors

Each type of telescope has its own particular advantages and disadvantages. All very large telescopes are of the reflecting type. This is because it is impossible to construct lenses of large size, since a large volume of flawless glass is needed, and a gigantic lens would bend under its own weight. A lens can only be supported round its rim, whereas a mirror can be supported completely underneath, to prevent bending.

24. The Anglo-Australian telescope at Siding Spring, 400 km north west of Sydney, NSW. The main mirror has a diameter of 3·9 m. (Anglo-Australian Observatory)

For small telescopes, refractors often perform better than reflectors of the same aperture. One reason for this is that less light is lost on transmission through two or three thin lenses than by reflexion at several mirrors, which easily become tarnished. Mirrors are usually made by coating a glass surface with a thin film of aluminium. Lenses, however, suffer more than mirrors from optical faults or **aberrations**. The chief one of these is the false colouring around images caused by the lens acting to a certain extent as a prism, and breaking up the light into its component colours. This effect is known as **chromatic aberration**.

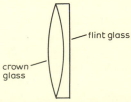

26.5 The construction of an achromatic doublet

Chromatic aberration can be reduced by constructing an objective lens from two or more parts, each made from a different type of glass. The various components counteract the dispersion produced by the other lenses; commonly two lenses, one each of crown and flint glass are used and this composite lens is called an **achromatic doublet**.

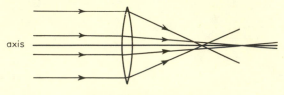

26.6 Spherical aberration

Lenses and spherical mirrors also suffer from **spherical aberration**. Rays of light that strike the edges of the lens or mirror are brought to a focus at a different place from those that strike the middle. Spherical aberration in lenses can be reduced by the use of a suitably designed doublet lens which compensates for chromatic aberration at the same time. For mirrors, the shape of a paraboloid rather than part of a sphere is used. A paraboloid bring all rays to a focus at one point.

Telescope mountings

Most telescopes are mounted in such a way that they are free to point in any direction, and so to follow the motion of stars through the sky. To achieve this freedom the telescope needs to move about two independent axes, which are perpendicular to each other. If one of these axes is parallel to the Earth's axis of rotation, the telescope needs only to be moved around this axis to keep track of the apparent motion of the stars. A mount like this is called an **equatorial mount**. This method is used for nearly all larger telescopes, although the exact design may vary in detail. An electric motor can be used to drive the telescope at a rate which compensates for the motion of the stars. The axes are often equipped with engraved scales (called setting circles) so that an object can be found by setting on its right ascension and declination.

An equatorial mount has to be specially constructed for the latitude of the observatory, as the polar axis has to point directly to the north pole of the sky, whose altitude, of course, varies with latitude. Equatorial mounts are not, therefore, particularly suitable for portable telescopes, which usually have an **alt-azimuth mounting** on a tripod. In this arrangement the telescope can rotate parallel to the horizon, and also in altitude. The disadvantages are that varying degrees of motion around both axes are

25. The large radio dish of the Max-Planck Institute for radio astronomy, Bonn. The dish is 100 m in diameter and fully steerable. (Max-Planck Institute)

required to follow the stars, and the field of view rotates as the telescope is moved. This system of mounting can only be used for a large telescope if there is a computer to control the varying rate at which it must be driven.

Radio telescopes

During the last few decades radio astronomy has become increasingly important. In certain ways, radio telescopes follow the same principles of construction as optical ones, but since one cannot see radio waves, electronic equipment has to be used for detecting them.

Radio signals from space are very weak. Consequently, a large reflector is used to collect as much radio energy as possible. The waves are focused on to an aerial where they create an electric signal which is fed into a radio receiver and amplified before it is recorded by, for example, a pen-recording voltmeter or immediately analysed by computer.

There is another good reason for using a large reflector. The resolution of a telescope depends not only on the aperture size, but also on the wavelength of the radiation. As the wavelengths of radio waves are greater than the wavelengths of light by a factor of about a million, very large dishes are needed, and even these give poor resolution compared with optical telescopes. However, the resolution problem is mainly overcome by the technique of linking up several smaller dishes, some distance apart, in an arrangement known as an **interferometer**. This set-up gives the same resolution as a single dish with a diameter as large as the separation of the small dishes. Using such instruments it is possible to make radio maps of the sky.

Astronomy from space

A telescope located on the surface of the Earth suffers from several disadvantages: (i) the Earth's atmosphere absorbs radiation in the infrared, ultraviolet, and X-ray regions of the spectrum; (ii) the Earth's atmosphere is itself a powerful source of infrared rays; (iii) currents of air distort the images obtained by optical telescopes; (iv) almost all optical observatories throughout the world are now seriously threatened by scattered light from cities and large towns. To overcome these difficulties techniques for carrying out observations in space have been developed. Infrared astronomy, for example, has been carried out by telescopes flown to a very high altitude in specially-adapted aircraft. Ultraviolet and X-ray observations are made by automatic satellites that are orbiting the Earth as well as by detectors launched on rockets. In the case of satellites, the telescopes are controlled remotely by instructions radioed from the Earth, and the measurements are radioed back by the satellite. Strong X-ray stars and powerful X-ray galaxies have been discovered by the use of satellites. Optical astronomers hope that they will benefit from the perfect observing conditions in outer space when the Space Telescope is put into orbit round the Earth in the 1980s, and when permanent astronomical observatories are operating in space by the end of the century.

27 Cosmology: the nature of the Universe

General principles

The study of the Universe as an entirety is called cosmology. It may be conveniently divided into two parts: observational cosmology, which deals with the observations that tell us about the general properties of our Universe, and theoretical cosmology, in which mathematics is used to describe the overall appearance of the Universe. Cosmology is intrinsically difficult because of the uniqueness of the object being studied — the Universe — and the impossibility of performing experiments. We just have to make what we can of what the Universe is prepared to tell us! In order to make the subject at all manageable it is customary to make certain assumptions, as follows:

  (i) on the average the Universe is the same wherever we look at it.

  (ii) the laws of physics are the same throughout the Universe.

Not all cosmologists agree with these assumptions; however, the subject becomes very complicated without them. Assumption (i) means that there are no privileged observers: on a sufficiently large scale and for a set point in time the Universe will look more or less the same in terms of its density, temperature, and so on, no matter where we go. Assumption (ii) implies that constants such as the speed of light have the same value everywhere, that laws such as the Law of Gravitation are the same everywhere, and that neither the constants nor the laws are changing with time.

Observational cosmology

The dark night sky provides us with an important observational fact about our Universe: the sky is dark except where there are stars. To see that this is important suppose that the Universe stretched out for ever and ever, and that it were uniformly populated with stars and galaxies. The light intensity received from any one star or galaxy will be inversely proportional to the square of its distance from us, so the faraway galaxies contribute little light on an individual basis. But the total number of visible galaxies should increase as the cube of the distance, since the volume of space surveyed increases as the cube of the distance. Therefore, if we add up the total light received from all the galaxies in an infinite and stationary Universe we find it is infinity! The dimming of the light from remote (and therefore faint) galaxies by the distance is exactly cancelled out by the much larger number of galaxies in the bigger volume of space far away. Therefore, if our argument is right, the night sky should have a blinding intensity. This absurd conclusion is called Olbers' Paradox. The fact that the sky *is* dark at night shows that an assumption is wrong. Our mistake is that we left out the fact that the Universe is expanding: faraway parts of the Universe send us redshifted light, and this has lost some of its energy over and above the inverse-square law dimming. The dark night sky, therefore, tells us that we live in an expanding Universe.

Radio astronomy has provided us with vital clues about the nature of the Universe. Radio galaxies and quasars can be detected by radio telescopes even though they are at very great distances, perhaps as much as 10,000 million light years or more, from us. They indicate to us what the Universe was like long ago. Since they are the most distant known objects they can teach us about the properties of the Universe on the largest scale. It has been found that the density of radio sources increases with distance from the Milky Way at first, and then, at very great distances (beyond the reach of optical telescopes) the density declines. This tells us that the density of radio sources in the Universe varies with time; not so long ago there were more than there are now, but in the early Universe, which is what we are looking at when we detect objects 10,000 million light years away, there were fewer. Therefore, it appears that the nature of the Universe is changing as the Universe ages.

Radio astronomers have also detected a faint background of radiation that is uniform over the entire sky. It matches the continuum spectrum of a body with a temperature of $2.7°$ K. This uniform signal is believed to be the weak relic of an earlier phase of the Universe's history, when it was much hotter and denser than it is now.

We can ask ourselves if the Universe looks more or less the same in every direction. The answer is that it does appear to have the same properties everywhere, because no matter where we look:

(i) the density of galaxies and galaxy clusters is uniform
(ii) the background radiation is uniform
(iii) the distant radio sources are uniformly distributed across the sky.

Theoretical cosmology

It is the task of theoretical cosmology to explain why the Universe looks like it does.

Some years ago the Steady State Theory of the Universe enjoyed popularity. Essentially, this theory stated quite simply that 'things are as they are now because of how they were in the past'. The theory proposed that the Universe had existed for an infinite amount of time, that it would continue to exist for all time, and that it looked exactly the same everywhere. The expansion of the Universe causes the galaxies to get further apart all the time. To overcome the obvious difficulty that the density of the Universe then changes with time the Steady State cosmologists proposed that new matter appeared spontaneously to fill the gaps between the expanding galaxies.

The discoveries that the density of radio sources apparently varies with distance and that the Universe is filled with weak radiation are thought to be in disagreement with the Steady State theory by most cosmologists.

An alternative model is that of the Big Bang Universe. In this theory it is supposed that long ago the Universe was very small, extremely dense, and tremendously hot. From this state (sometimes called the primeval atom) the Universe has expanded steadily, and its properties have changed as it has done so. This accounts for the variation in the number of radio sources. The background radiation is the dying echo of the early very hot phase. The initial explosion was responsible for the observed recession of the galaxies. The Big Bang cosmology gives a reasonable account of the observed properties of the Universe.

Among the unanswered questions in cosmology are these:

(i) what preceded the Big Bang?

(ii) will the Universe expand forever?

(iii) does the Universe contain a significant amount of invisible matter?

The resolution of these fundamental problems is a challenging goal for astronomy and cosmology.

Appendix A

Index notation

Index notation is a shorthand method of writing very large and very small numbers, which avoids the use of strings of zeros. It is best explained by a number of examples:

$$10^2 = 10 \times 10 = 100$$
$$10^3 = 10 \times 10 \times 10 = 1000$$
$$10^4 = 10 \times 10 \times 10 \times 10 = 10000$$
$$\text{etc.}$$

$$3000 = 3 \times 1000 = 3 \times 10^3$$
$$3500 = 3 \cdot 5 \times 1000 = 3 \cdot 5 \times 10^3$$

$$10^{-1} = 1/10 = 0 \cdot 1$$
$$10^{-2} = 1/100 = 0 \cdot 01$$
$$10^{-3} = 1/1000 = 0 \cdot 001$$
$$\text{etc.}$$

$$0 \cdot 003 = 3 \times 0 \cdot 001 = 3 \times 10^{-3}$$
$$0 \cdot 0035 = 3 \cdot 5 \times 0 \cdot 001 = 3 \cdot 5 \times 10^{-3}$$

Appendix B

Units and abbreviations

The following standard units and abbreviations are used:

length:	m	metre
	km	kilometre $= 1000$ m
	mm	millimetre $= 0.001$ m
	cm	centimetre $= 0.01$ m
	nm	nanometre $= 10^{-9}$ m

mass:	g	gram
	kg	kilogram $= 1000$ g
	tonne	$= 1000$ kg (metric ton)

time:	yr	year
	d	day
	h	hour
	m	minute
	s	second

It is always clear from the context whether m is being used for metres or minutes.

The following special units are defined in the text:

A.U.	Astronomical Unit	page 13
Å	Ångstrom	page 82
pc	parsec	page 90

1 Mpc $= 1$ megaparsec $= 10^6$ pc

Temperature

Both the Celsius (°C) and Kelvin (°K) scales are used according to which is most appropriate. In both scales the size of the degree is the same, but in the Celsius scale the zero point is the freezing point of water, while on the Kelvin scale zero is absolute zero.

$$0°K = -273°C$$
$$273°K = 0°C$$
$$373°K = 100°C$$

Appendix C

The Greek alphabet

α	alpha	ν	nu
β	beta	ξ	xi
γ	gamma	o	omicron
δ	delta	π	pi
ε	epsilon	ρ	rho
ζ	zeta	σ	sigma
η	eta	τ	tau
θ	theta	υ	upsilon
ι	iota	ϕ	phi
κ	kappa	χ	chi
λ	lambda	ψ	psi
μ	mu	ω	omega

Appendix D

Conversion factors for units of measurement

| Length | 1 kilometre (km) = 0.621 miles |
| | 1 metre (m) = 39.37 inches |

Length   1 kilometre (km) = 0.621 miles
                        1 metre (m) = 39.37 inches

Mass   1 kilogram (kg) = 2.205 pounds
                        1 tonne = 1000 kg
                        1 tonne = 2205 pounds

Temperature $0°C$ = freezing point of water = $32°F$
                    $100°C$ = boiling point of water = $212°F$

Suggestions for further reading and reference

This list is by no means exhaustive and many other equally good books will be available in your library and bookshops. An illustrated booklet on astronomy, which is available for a modest price from the Royal Astronomical Society, Burlington House, Piccadilly, London W1V 0NL, includes a more extensive book list.

1 General Astronomy, but at a more advanced level than this book:

Survey of the Universe by D. H. Menzel, F. L. Whipple and G. de Vaucouleurs (Prentice-Hall, 1970)

Modern Astronomy by L. Oster (Holden-Day, 1973)

New Horizons in Astronomy by J. C. Brandt and S. P. Maran (Freeman and Co, 1972)

The Cambridge Encyclopaedia of Astronomy edited by S. Mitton (Jonathan Cape, 1977)

2 Practical astronomy for amateurs:

Astronomy with Binoculars by James Muirden (Faber and Faber, 1976)

Astronomical Photography at the Telescope by Thomas Rackham (Faber and Faber, 3rd edition, 1972)

Naked Eye Astronomy by Patrick Moore (Lutterworth Press, 4th edition, 1976)

3 Astronomical Maps and Data:

Norton's Star Atlas by A. P. Norton (16th edition extensively revised) (Gall and Inglis, 1973)

Constellations by J. Klepešta and A. Rükl (Hamlyn, 1969)

Astronomy Data Book by J. Hedley Robinson (David and Charles, 1972)

Philip's Planisphere. This is a sturdy plastic map which is simple to use. The stars visible at any date and time can be set on a dial. Generally available from map shops and some book shops.

4 'O' level Astronomy:

For details of the current syllabus and method of entry, write to:

University of London School Examination Department, 66–72 Gower Street, London WC1.

5 Astronomical Societies

National Societies:

The Junior Astronomical Society, 58 Vaughan Gardens, Ilford, Essex.

The British Astronomical Association, Burlington House, Piccadilly, London W1V 0NL.

The BAA publishes annually a Handbook full of invaluable data for the amateur observer. Write to the Secretary for details of publications and membership.

A list of local astronomical societies in Great Britain is printed in the Yearbook of Astronomy, edited by Patrick Moore and published in about November each year by Sidgwick and Jackson.

Index

Abell 1060, 120
aberrations, of optical systems, 125–126
Adams, John, 42
Adonis, 41
Airy, G. B., 42
Algol, 103
alt-azimuth mounting, 126
altitude, 60, 62
Ångström, 82
Apollo, 41
Apollo 15, 18, 73
association, stellar, 112
asteroids, 41
Astronomical Unit, 13
atomic mass number, 84
atomic number, 84
atoms, 83
aurorae, 46–47
azimuth, 60, 62

Barnard's star, 91
Betelgeuse, 107, 112
Big Bang theory, 131
binary stars, 101–104
black hole, 100
Brahé, Tycho, 50
Butterfly diagram, solar, 78

Callisto, 37
Cassegrain telescope, 123
Cassini division, 39, 40
celestial equator, 59, 64
celestial sphere, 58
Cepheid variables, 105–106, 119
Ceres, 41
Challis, James, 42
circumpolar stars, 60, 61
clusters, of galaxies, 119, 120

clusters, of stars, 97, 112–114
Coal Sack, 108
comets, 48–52
compound, 83
constellations, 62
corona, solar, 79
cosmology, 129–131
coudé telescope system, 123
Crab Nebula, 98–99
craters, Mars, 32
   Mercury, 28–29
   Moon, 16, 18
   Venus, 31
culmination, 60

day, sidereal, 55
   solar, 55
declination, 58
degenerate matter, 95
Deimos, 34
dichotomy, 27
dispersion, 82
doppler effect, 86
dwarfs (stars), 92, 94

Earth, 44–47, 55–60
eclipse, annular, 23–24
   lunar, 24, 26
   solar, 23–26
   of Jupiter's satellites, 37
eclipsing binary stars, 102
ecliptic, 64
electromagnetic radiation, 81–83
electrons, 83
elements (chemical), 83
elongation, of Mercury and Venus, 27
Encke's Comet, 50
equation of time, 68

equatorial mount, 126
equinoxes, 65
Eros, 41
escape velocity, 73
Europa, 37

first point of Aries, 59
Flamsteed, 63
flares, solar, 75

galactic clusters, 112–113
galaxies, 115–121, 129–130
Galaxy (Milky Way), 14, 108–114,
    115, 119
Galileo, 37
Galle, Dr., 42
gamma-ray astronomy, 83
Ganymede, 37
giants (stars), 92, 94, 95, 97, 98, 106,
    107
globular clusters, 112–114
granulation, solar, 74
gravitational field, 72, 73
Great Red Spot, 35, 36
greenhouse effect, 30
Greenwich Mean Time, 68

h and χ Persei, 113
Halley, Edmund, 50
Halley's Comet, 50, 51, 54
Herschel, W., 41
Hertzsprung-Russell diagram, 94–95,
    97, 113
Hidalgo, 41
Horsehead Nebula, 110
hour angle, 69
Hubble, E., 115, 119
    —Law, 119, 121
Hussey, Rev. T. J., 42
Hyades, 113

inferior planets, 27
infrared astronomy, 83, 128
interferometer, stellar, 92
    radio, 128
interstellar matter, 108–112
Io, 37, 38
isotopes, 84

Juno, 41
Jupiter, 35–38

Kepler, Johannes, 98
    laws of planetary motion, 70, 103

Leavitt, Miss H., 105
Le Verrier, 42
libration, 21
limb darkening, 74
Lowell, Percival, 42

M5, 113
M13, 113
M31, 14, 119
M33, 119
M42, 111–112
M82, 118
M83, 116
Magellanic Clouds, 105, 119
magnification, of telescope, 123
magnitude, 91
main sequence, 94
Mariner 9, 32
Mariner 10, 28, 29
Mars, 32–34
mass-luminosity relation, 92
Mercury, 27–29
meridian, 60
Messier, 14
meteorites, 53
meteors, 53–54
meteor showers, 53–54
Milky Way—see Galaxy
minor planets, 41
Mira Ceti, 106
molecules, 83
    in stars, 88
    interstellar, 112
month, sidereal, 20, 66
    synodic, 20
Moon, 16–22
    path of, 66

nadir, 60
nebulae, 99, 109, 111, 115
Neptune, 42, 43
neutrons, 83

neutron star, 98–100
Newtonian telescope, 123
Newton's law of gravitation, 50, 71
  laws of motion, 71, 73
NGC 1365, 117
nova, 107
nucleus, atomic, 83

occultation, by Moon, 21
  of Jupiter's satellites, 37–38
Olbers' Paradox, 130
Omega Centauri, 113–114
open clusters, 109, 112–113
opposition, of Mars, 32
  of planets, 67
Orion nebula, 111–112

Pallas, 41
parallax, lunar, 22
  stellar, 90
parsec, 90
Phobos, 34
photons, 81
photosphere, solar, 74
Piazzi, 41
Pickering, E. C., 43
Pioneer 10, 35, 36
plages, solar, 75
Pleiades, 109, 111–113
Pluto, 42–43
Polaris, 58
precession, 56–57, 66
Praesepe, 113
prominences, solar, 75, 77
proper motion, 91
proton-proton chain, 96
protons, 83
protostar, 95
Proxima Centauri, 14, 90
pulsar, 99

quasar, 121, 130

radial velocity, 91
  of spectroscopic binary, 102
  of galaxies, 119
radio astronomy, 83, 112, 127–128,
  130

radio telescope, 127–128
red giants, 95, 97, 98, 106, 107
redshift, 121
resolution, of telescope, 124
retrograde motion, of planets, 66–67
right ascension, 58, 59, 66
RR Lyrae stars, 106

Saturn, 39–40
Saturn's rings, 39–40
seasons, 55, 56
shooting star, 53
sidereal day, 55
sidereal month, 20, 66
sidereal time, 68–69
solar activity, 74, 78–79
solar cycle, 78
solar day, 55
Solar System, 14, 50, 51, 52
solar time, 68
solar wind, 50, 79
solstices, 65
spectroscopic binary stars, 101
spectrum, 82
  line, 84
  continuous, 85
  stellar, 87–89
standard time, 68
stars, binary, 101–104
  circumpolar, 60–61
  composition, 89
  clusters, 97, 109, 112–114
  distances, 90, 106
  evolution, 95–100
  masses, 92–93
  names, 62
  nearest, 15
  radii, 92
  variable, 98, 105–107
Steady State theory, 130–131
Stefan-Boltzmann Law, 86
stellar evolution, 95–100
Sun, 74–80
  path of, 64–65
  properties of, 80
sunspots, 74, 75, 78
supergiants, 95, 106, 107
supernova, 98–99, 107

synodic month, 20
synodic period, of a planet, 67

telescope, 122–128
terminator, 20
time, 68–69
Titan, 40
Tombaugh, C., 43
transit, of Mercury and Venus, 28
    of Jupiter's satellites, 37
    of meridian by stars, 60
47 Tucanae, 113

ultraviolet astronomy, 128
Universal time, 68
Uranus, 41–43
Uranus' rings, 43

Van Allen belts, 46–47
variable stars, 105–107
Venera 9 and 10, 31
Venus, 27, 28, 30–31
Vernal equinox, 59, 66
Vesta, 41
Viking 1 and 2, 32–34
Visual binary stars, 101

white dwarfs, 95

X-ray astronomy, 83, 100, 128

year, 55

zenith, 60
zodiac, 64